THE MINERAL BOOK

DAVID MCQUEEN

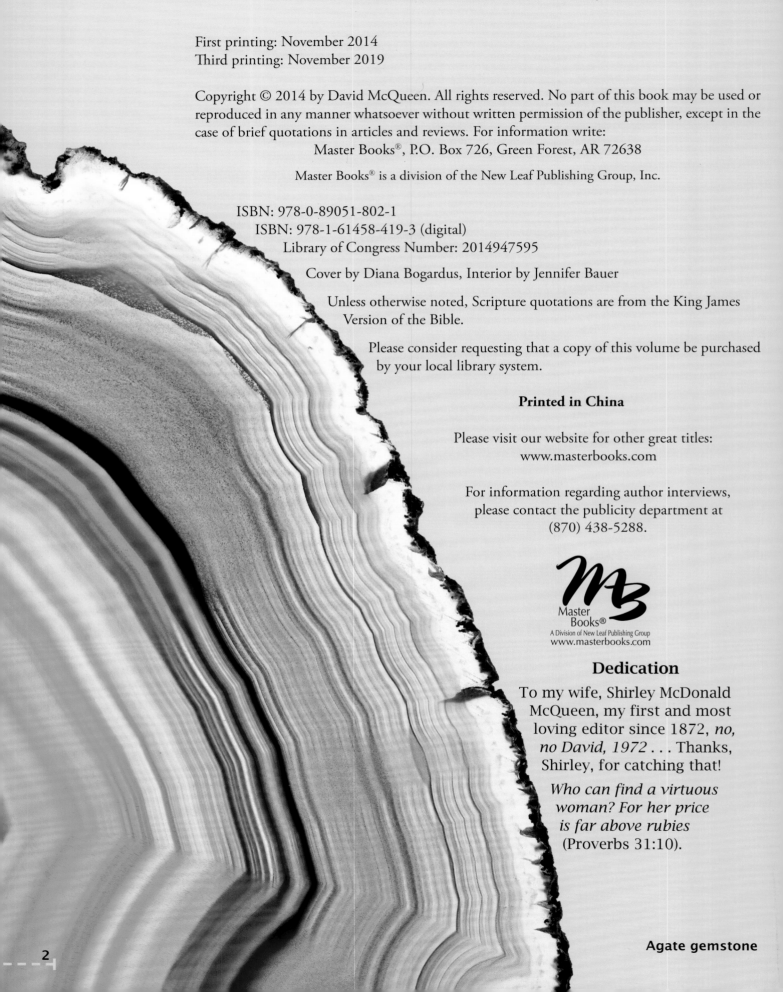

First printing: November 2014
Third printing: November 2019

Master Books® is a division of the New Leaf Publishing Group, Inc.

ISBN: 978-0-89051-802-1
ISBN: 978-1-61458-419-3 (digital)
Library of Congress Number: 2014947595

Cover by Diana Bogardus, Interior by Jennifer Bauer

Unless otherwise noted, Scripture quotations are from the King James Version of the Bible.

Please consider requesting that a copy of this volume be purchased by your local library system.

Printed in China

Please visit our website for other great titles:
www.masterbooks.com

For information regarding author interviews, please contact the publicity department at (870) 438-5288.

Master Books®
A Division of New Leaf Publishing Group
www.masterbooks.com

Dedication

To my wife, Shirley McDonald McQueen, my first and most loving editor since 1872, *no, no David, 1972* . . . Thanks, Shirley, for catching that!

Who can find a virtuous woman? For her price is far above rubies (Proverbs 31:10).

Agate gemstone

Calcite under ultraviolet light.

TABLE OF CONTENTS

Cover image: Large symmetrical piles of salt (halite)

Fluorite

Foreword

Minerals continue to fascinate me. They have such beauty, such elegance—particularly when viewed through a microscope or in their crystal phases. And they also contain a historical message, especially when considered from a biblical creation perspective. They tell us much about the world's origin and the timing of events mentioned in Scripture. From minerals we can learn about creation itself, the great Flood of Noah's day, and even details pertaining to the history of Israel.

Mr. Dave McQueen has been my dear friend for many years. As geologist colleagues we have traveled far and wide on various geological field trips. Even though our geological work together has most often centered on the rock strata or the fossils within the rocks, Dave always considers the minerals, too. Before working with him, I had seldom understood their full importance.

In the pages to follow, Dave will open up the world of minerals to you, just as he did for me. You will thrill with the surpassing design of crystals and the microscopic detail of transparent thin sections of minerals. A whole new aspect of God's wondrous creation will be revealed before you. You will grow to appreciate the monetary value of some minerals and learn how creation scientists use other tiny minerals in creation research to expand our understanding of both our Creator and His marvelous creation, and our place in it.

So enjoy your educational excursion into the fascinating world of minerals. Allow them to not only increase your understanding of the physical world around you but deepen your understanding of its Creator as well.

John Morris
President of the Institute for Creation Research

Our best-selling
Wonders of Creation Series
is even better!

The series is being developed with an enhanced educational format and integrated with a unique color-coded, multi-skill level design to allow ease of teaching the content to three distinct levels.

SAFETY FIRST !

This book contains encouragements for field work and mineral collecting on roadsides. For your safety, we ask that you always have adult supervision.

How to Use This Book

The Three Skill Levels

The Mineral Book has been developed with three educational levels in mind. These can be utilized for the classroom, independent study, or homeschool setting. For best possible comprehension, it is recommended that every reader examine the text on the gray background. More skilled readers can then proceed to the green sections as well. Finally, the most advanced readers may read through all three sections. Look for the following icons and special features throughout the book:

| Level 1 | Level 2 | Level 3 |

Level One

Introduces minerals to younger readers. The information written for this level serves to whet the appetite for the naturally inquisitive child's mind. The young reader will be drawn to the colorful minerals they have already discovered in nature, in their classrooms, and in museums. It points them to God's creative design.

Level Two

Written for those with a more expanded vocabulary, and introduces them to mineral identification and to minerals that are mentioned in the Bible. This age will be challenged to begin collecting minerals from their environment or as they are vacationing. As they read this book, they will be motivated to collect and study minerals with a whole new perspective on how important these have been, and continue to be, to our world.

Level Three

Written at the highest skill level, and will especially be of value to amateur mineral collectors, who want to make more sense out of their beautiful collections. It is our earnest desire that level-three discussions inspire young men and women to become creation scientists. This information will serve to have them look at minerals from a biblical, scientific viewpoint of these God-designed gifts to us.

Chapter Mineral Focus

At the beginning of each chapter is a level-two introduction of a specific mineral, detailing its name, chemical formula, crystal system, hardness, luster, and streak, as well as a biblical passage that relates to the mineral (Rock Solid Minerals). Also, details are included concerning where this particular mineral is found and what it is used for. The Fun Fact on the page is always level-one content.

ROCK SOLID
IT'S IN THE WORD
MINERALS

The Scarlet Thread and the Colored Stones

Dr. W.A. Criswell did a detailed study on the significance of the blood sacrifice that God has required for redemption. He imagines the references to the blood to be like a scarlet thread that begins in Genesis, when God had to kill the first animal in order to clothe Adam and Eve after sin shattered their innocence. This scarlet thread winds through Leviticus, when the priest performed animal sacrifices to cover the sin of God's people, and culminates in the New Testament, where Christ becomes the sacrificial lamb and is slain, so that His blood could cover our sins.

Dr. Criswell refers to a physical scarlet thread in Joshua 2:18. Joshua has sent spies into the city of Jericho, which God had commanded them to destroy. The spies were hidden by a harlot named Rahab. God had already been working in her heart, because she risked her life for these men. She had heard of these people who served a God who struck fear in the hearts of their enemies. She wanted to be on their side! Before they escaped the city, she knew they would be back to destroy it. She asked them to show her a kindness and spare her and her family. In Joshua 2:18, the spies answered her: "Behold, when we come into the land, thou shalt bind this line of scarlet thread in the window which thou didst let us down by: and thou shalt bring thy father and thy mother, and thy brethren, and all thy father's household, home unto thee."

What a wonderful analogy of being saved by a scarlet thread to that of being saved by the red blood of Jesus Christ. Rahab was a harlot, yet her name is included in the genealogy of Christ! (Matthew 1:1–16). One person's faith has a huge impact on historical outcomes.

Creator. Savior. Lord.

These three names are represented by the three stones shown throughout in this book. It helps us tie together the idea that God has loved us eternally. He created a beautiful world and then created us in His own image, (purple stone of creation), He planned a way to save us (red stone of the blood), and in sending His only Son, He did redeem us and offer us His lordship (gold nugget of His lordship in our lives.)

Criswell, W.A., 1979, *The Criswell Study Bible* (KJV): Nashville, Thomas Nelson, approx. 1,600 p. Note p. 270 (comments on Joshua 2:18-24)

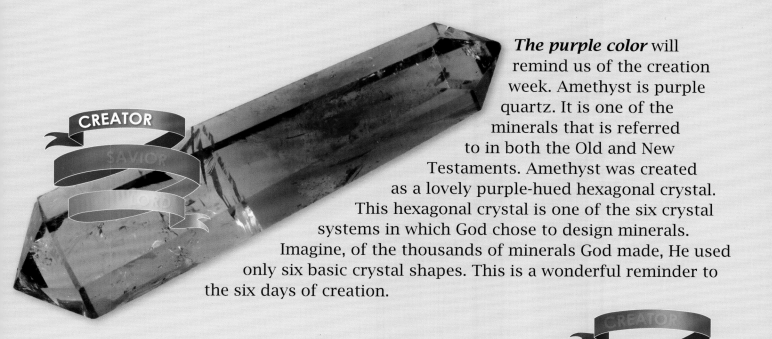

The purple color will remind us of the creation week. Amethyst is purple quartz. It is one of the minerals that is referred to in both the Old and New Testaments. Amethyst was created as a lovely purple-hued hexagonal crystal. This hexagonal crystal is one of the six crystal systems in which God chose to design minerals. Imagine, of the thousands of minerals God made, He used only six basic crystal shapes. This is a wonderful reminder to the six days of creation.

The red color will remind us of Jesus our Savior. The red emphasizes the blood of Christ, which is God's gift of grace to us. God created jasper as one of the beautiful red minerals mentioned in both the Old and New Testaments. As you read about the many minerals God has given us, and learn how they allow us to live a richer life, I think you will begin to understand that they are, just as grace is, something we get from God that is totally undeserved.

The gold color will remind us of Christ's lordship in our lives. When we put our trust in Christ, we crown Him as King of our life. **Gold** is mentioned at least 350 times in the Bible. It is considered one of the most precious minerals. Gold is a soft metal, which is easily shaped, or malleable. Think of Christ's work in our hearts, as we read His word, and imagine that your heart, which is God's greatest treasure, to be like gold in His hand. He is shaping it to know Him, and worship Him as our Creator. Psalm 100 tells us that it is God who has made us, not we ourselves. **YOU** are His creation. This speaks of our relationship to Him, which will last throughout all eternity!

CHAPTER 1

Where Do We Find Minerals?

There are an estimated 5,000 minerals in the world. They can be found in every locality around the globe. Minerals appear in caves, in deserts, on ocean floors, in mountain ranges, and in river sediments, just to name a few. Some are native minerals (also called native elements), such as gold, copper, and silver, and contain only one element. The common mineral quartz is found in beach sand and river sediment that is scooped out to be part of the mixture we call concrete.

The Bible is full of references to minerals, rocks, metals, and gems. Gold is mentioned as early as in Genesis, the first book of the Bible. King Solomon was known throughout the ancient world for his mineral treasures. The book of 2 Chronicles records what is called "a covenant of salt" that was used as a sign of an unbreakable agreement, and in Matthew 26, our Savior was betrayed by Judas for 30 pieces of silver. These and other minerals are found all around us.

Terms

Rare minerals – *Those minerals that are more uncommon, generally more valuable, and often harder to gather because of the process involved.*

Chemical interactions – *The reaction that occurs when two or more chemicals are combined.*

Native mineral – *A native element (or mineral) you pick up that looks like a rock, but is actually a mineral composed mostly of a metal.*

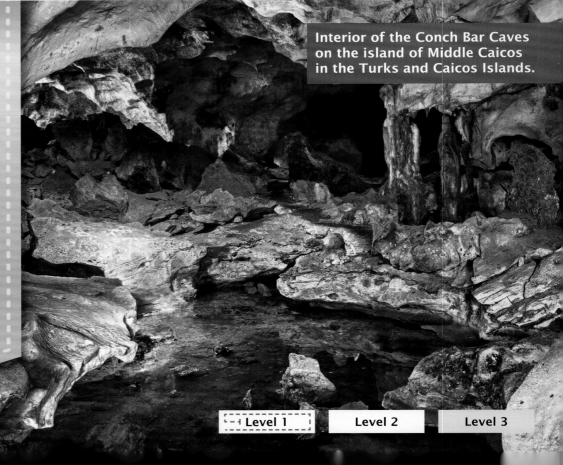

Interior of the Conch Bar Caves on the island of Middle Caicos in the Turks and Caicos Islands.

Level 1 | Level 2 | Level 3

MINERAL FOCUS	Salt or halite
CHEMICAL FORMULA	NaCl
CRYSTAL SYSTEM	Cubic
HARDNESS	2½
LUSTER	Vitreous
STREAK	White

ROCK SOLID MINERALS

IT'S IN THE WORD

Matthew 5:13

Ye are the salt of the earth: but if the salt have lost his savour, wherewith shall it be salted? it is thenceforth good for nothing, but to be cast out, and to be trodden under foot of men.

Where is it found? Salt is found in mines, worldwide. Sea salt is evaporated from ocean water. The largest salt mine in the United States is in New York. The largest salt mine in the world is in Ontario, Canada.

How is it used? Salt is essential to our nutrition. It is the sodium ions present in salt that the body requires in order to perform a variety of functions. Sodium (Na) helps maintain the fluid in our blood cells and is used to transmit information in our nerves and muscles. It aids in digestion by the intake of certain nutrients from our small intestines. Salt is also used to preserve meats and vegetables. Salt melts ice, so it is used in cold climates to clear the roads.

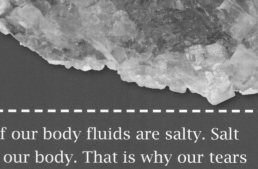

Fun Fact: All of our body fluids are salty. Salt is in every cell in our body. That is why our tears and our sweat taste salty!

Salt Production: Methods and Miners

Salt is a vital mineral needed by people around the world. The process of producing salt is generally done in one of the following three ways:

Evaporation: In drier coastal climates, near salty sea water or salt lakes, salt water is directed into shallow pools, where the wind and sun help evaporate the water and leave behind the salt.

Deep-shaft mining: Much like regular mining for minerals like zinc and copper, this involves drilling shafts into the earth where salt deposits are found, crushing the salt, and bringing it to the surface, usually to be used as rock salt.

Solution mining: Most table salt is made from gathering salt from salt beds and injecting water into the mix to remove the salt. This brine solution is then evaporated in salt pans at a processing plant.

The workers involved with the solution mining or salt panning are often exposed to very harsh conditions for very little pay. Both their exposure to the salty brine with little protective gear and exposure to harsh weather conditions takes its toll on their health, and the livelihood of their families, who often must travel with them wherever the work can be found.

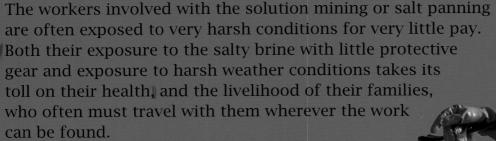

Salt pans in India dry up sea water to produce salt.

Salt pans in Maras, Peru, have been in use since the Inca culture, approximately 500 years ago.

How important are minerals to us?

Let's just imagine a typical morning. Before you taste your eggs, you salt them. You have just added a mineral. Salt is NaCl, an extremely important mineral in our diet. Picking up a fork, you eat your eggs. This stainless steel utensil is made by mixing iron with another metal called chromium, which stops the fork from rusting. Who wants to eat from a rusty fork? The iron and chromium, by the way, both come from minerals! As you fill your glass with milk, you may be surprised to know that your glass started out as pure quartz sand. The quartz (SiO_2) was melted down and mixed with other ingredients to make it transparent and leak-proof! Next you pick up a pencil with a lead made from graphite, a soft mineral that enables you to write a quick note to your dad, reminding him of your dental appointment after school. You quickly stick this to the wall, which is drywall, which is made from heated gypsum plaster. Did you know gypsum is a mineral? Grabbing your backpack, you head out the door to catch your ride. You walk down a brick path that leads to the driveway. The bricks are made from clay, a very common and plentiful mineral. There you have it. You have benefited from at least seven minerals, and the day has just begun!

The deep waters of the Dead Sea in Israel are shown dark blue, while brighter blues indicate shallow waters or salt ponds. In the modern age, sodium chloride and potassium salts evaporated from the sea are used for water conditioning, road deicing, and the manufacturing of polyvinyl chloride (PVC) plastics. The expansion of massive salt evaporation projects are clearly visible over the span of 39 years.

Minerals in the Human Body

There are a good number of minerals beyond just salt in the human body that are vital for our health, and are the foundational material for our bones, cells, and tissues. Your body can't produce the minerals it needs, so God has provided sources all around us to provide you with the balance you need. Here are just a few of the needed minerals, where you can get them, and what they do for your body!

Enzymes in your body, more than 200 of them, need zinc to help them process the chemical reactions. Zinc can be obtained from eating eggs, fish, various meats, and wheat germ.

Connective tissues in your body need silicon to help them form stronger ligaments and tendons, and also in the growth of our bones. Silicon can be found in root vegetables, like carrots, whole grain bread, and cereals.

Blood transports oxygen through your body, and for this to function well it needs iron, which works best in combination with vitamin C. Your body can get iron from many sources, which include eggs, oysters, red meat, and seeds.

Cells need potassium and sodium to help them function, as well as to regulate the water in your body. A good source of potassium is citrus fruit, as well as nuts, leafy green vegetables, and potatoes, while sodium comes from table salt, fish, and other salted meats.

Bones contain masses of crystals composed of elements like calcium and phosphorus to make them stronger. Calcium can best be obtained from milk and other dairy products, as well as green vegetables and nuts. You can get phosphorus from chicken, eggs, fish, nuts, and seeds.

Mineral Composition

What is an example of a rare mineral? This question is not nearly as simple as you might think. Search online yourself. Some websites list apatite as a rare mineral, yet it is seen in every Mohs' mineral hardness set sold worldwide (see chapter 3). Why would anyone view apatite as an example of a rare mineral? Apatite's chemistry is not simple, yet not as complex as some. Here is what we mean: Apatite, which is rich in fluorine, is written as: $Ca_5(PO_4)_3F$. Apatite is actually a group of minerals. If it is rich in chlorine, the formula is written: $Ca_5(PO_4)_3Cl$. It is a phosphate mineral containing calcium, fluorine, chlorine, and hydroxide. Some scientists consider apatite rare because its particular crystal form is not frequently found.

Apatite

Are minerals an accident of chemical interactions, simply appearing on earth due to chance and impersonal forces of nature? Some secular books will tell you that minerals are naturally occurring substances, which implies that they are simply a part of mother earth. Instead, do minerals reflect an order and beauty that could only be designed by an intelligent Creator? This idea implies that each mineral type was specially created to reflect a God-designed universe. As you examine minerals, think of this verse: "For the invisible things of him from the creation of the world are clearly seen, being understood by the things that are made, even his eternal power and Godhead; so that they are without excuse" (Romans 1:20).

The more intensely one looks at the order and design of minerals, the more the truth of this verse in Romans rings true. The minerals we find on this amazing planet are lovingly designed by a God who wants us to see and believe that He exists! "But without faith it is impossible to please him: for he that cometh to God must believe that he is, and that he is a rewarder of them that diligently seek him" (Hebrews 11:6). As we look at minerals with this verse in mind, He will reward us with wisdom and awe that far surpasses the scientific minds of those who do not believe in the Creator God.

Minerals or Rocks?

It seems a bit confusing on the surface to say that minerals are not rocks, but all rocks are made up of minerals. This is why it is often said, "Let us get the minerals out of that rock!" Minerals contain metals. In mining, rocks are broken to yield minerals, which then are smelted to release metals such as gold or zinc. Geologists classify three types of rocks: igneous (like lava), metamorphic (like marble) and sedimentary (like sandstone). All three kinds of rocks were formed both during the Creation week and the Flood year.

Ancient Egyptian quarry

For thousands of years, people have mined minerals and other materials like coal in order to provide both aesthetic and useful benefits for their cultures. These mining operations have always been strenuous, as well as dangerous. In the mid-1800s, there was a movement to use young boys in coal mining; a movement that lasted nearly 60 years. They were called *breaker boys*, some as young as 8, and though the public often fought against the use of children in something so unsafe, they used them to mine in both the United States and the United Kingdom.

MINING

The breaker was where the coal was crushed, then sent to be sorted by the boys. They would often work 6 days a week, hard days lasting 10 hours or more. The work was exhausting, and they were liable to lose fingers if they couldn't pay attention, and their fingers were often cut by the sharp slate.

Nipper Willie Bryden at age 13.

Spraggers is what they called the boys who slowed the racing mine cars down with sprags, pieces of wood used as manual breaks. *Nippers* is what the boys were called who would open the doors when the mule and driver pulling the coal cars passed through, which meant nippers were often sitting alone in the damp darkness. *Mule drivers* were the older boys who helped bring down empty coal cars throughout the entire mine, and pulled out the loaded carts.

Thousands of young boys worked in the mines, their numbers diminishing as Christians applied the love of Christ to children. More and more child labor laws were created because of the public dismay over the conditions these children were forced to work in. Finally, by 1920 the use of breaker boys in the mines neared its end.

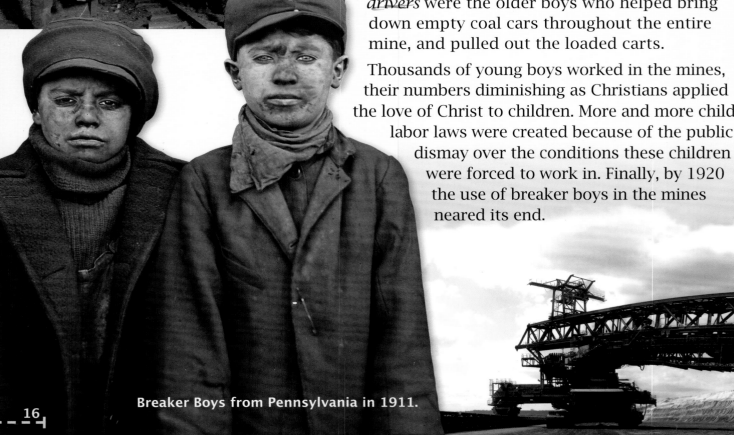
Breaker Boys from Pennsylvania in 1911.

Modern mineral mining operations are considerably different from mining in the past, but the dangers of mining still exist. Because of the value of the minerals, often operating a mine can run in the billions of dollars, including the expense of the equipment, the land rights, environmental regulations, trained personnel, transportation, time of processing, and more. And often the sites that hold these valuable minerals, such as gold or diamonds, are either in remote areas or involve deep and costly excavation procedures.

Sometimes the costs become too high to maintain operations if the return from minerals extracted is not as expected. One such site was the Beaulieu Mine in the Northwest Territories of Canada. It began production near the end of 1947, but only recovered about 30 troy ounces of gold by 1948. A troy ounce is used to measure precious metals and is slightly more than an ounce (1.09 ounces). The mine was closed soon after because of bankruptcy.

All the costs must be carefully weighed before new operations are presented to investors, and even then one can never know for sure about the return of investment possible. Mining feasibility studies are used to help companies and investors evaluate whether or not to proceed, but they are always limited. This is what has come of our modern need for certain minerals in a sinful world where they no longer simply fill the rivers, ready to be easily gathered by hand. We must work hard to collect them.

Marion steam shovel reminiscent of Mary Anne from the 1939 children's classic *Mike Mulligan and His Steam Shovel*.

Electric shovels can carry up to 56 cubic yards or 98 tons of ore in a single scoop. The trucks themselves cost about $3.5 million each!

The Bagger 288 is a bucketwheel excavator used in strip mining. It is also the largest land vehicle of all time.

17

As with many monuments and historic buildings, the Taj Mahal in India was made of various kinds of marble, beautifully carved with exquisite detail. Marble is a metamorphic rock and made up of minerals like calcite. For thousands of years, marble has been used by artists to form lifelike figures into sculptures. Carrara is a city in Italy that has supplied much of the beautiful white and blue-gray stone, used by artists including the marble Michelangelo used to sculpt his famous figure of Moses in Rome.

Minerals in Everyday Materials

A cloth towel: contains chromite and sphalerite for dyes.

Batteries: contain galena, graphite, and sphalerite.

Carpet: contains sphalerite (dyes), chromite (dyes), and sulfur (foam padding).

A computer: contains wolframite (monitor), copper (wiring), quartz (electronics), and silver.

Additional Research

Compare the definitions of minerals from three mineral or geology books in your library or from online sources. If you can, obtain a copy of *The Geology Book*[1] and read chapter 8. Note the differences between secular sources and Christian sources.

1 Morris, John D. *The Geology Book* (Green Forest, AR: Master Books), 2000.

The Wealth of Nations

Minerals and gems have played an important role throughout human history. Yet, man has no control over where the diamond or iron ore deposits are formed. From the biblical geology viewpoint, God decided where all minerals were placed. God's grace and mercy allowed people in the pre-Flood world and the post-Flood world to actually find mineral deposits necessary for refining metals that come from these deposits.

Wars have been fought over mineral wealth. In WWII, Germany invaded certain European areas because of their mineral wealth, in spite of the military risk. In addition, many of the art treasures of the world were painted using mineral-based pigments. Empires have been symbolized by golden crowns encrusted with precious gems. Most coins from ancient times until now have been made from metals, which are refined from minerals. Examples of these metals are gold, silver, copper, bronze, and brass.

The Dutch artist Johannes Vermeer (1632–1675) painted *The Milkmaid* using more expensive mineral–based paints, such as ultramarine from lapis lazuli, in order to bring out the vibrance of the colors.

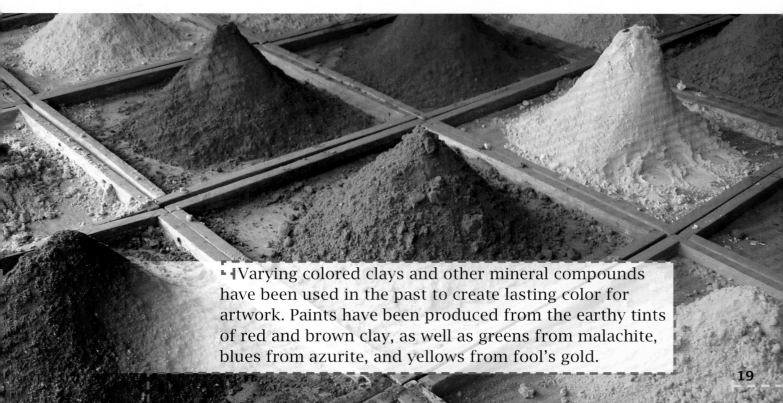

Varying colored clays and other mineral compounds have been used in the past to create lasting color for artwork. Paints have been produced from the earthy tints of red and brown clay, as well as greens from malachite, blues from azurite, and yellows from fool's gold.

2 What Is a Mineral?

Minerals are a wonderful gift from our Creator. He placed minerals in the Earth's crust for us to use. God, in His wisdom, knew that man would be able to find and use these minerals to build homes, print the Bible, season and preserve our food, paint beautiful pictures, manufacture everything from toasters to jet engines, and even adorn ourselves! How different our world would be without minerals.

But just what is a mineral? What makes it different from a rock or other chemicals? There are certain tests by which minerals are identified, and this chapter will help clarify how geologists begin to classify and order minerals on earth.

Terms

Geodes – *Spherical and hollow stones that contain quartz crystals inside them.*

Johannes Gutenberg (1395 — 1468) – *Found a way to use copper and other metals to construct a printing press.*

Cuprous – *Something that contains copper in the monovalent state.*

Cupric – *Something that contains copper in the bivalent state.*

Compounds – *Objects that contain two or more parts.*

Inorganic – *Containing no plant or animal material.*

Valence – *Determines the number of atoms that will unite in a chemical reaction.*

Level 1   Level 2   Level 3

MINERAL FOCUS	Amethyst
CHEMICAL FORMULA	SiO_2
CRYSTAL SYSTEM	Hexagonal/trigonal
HARDNESS	7
LUSTER	Vitreous
STREAK	White

ROCK SOLID
IT'S IN THE WORD
MINERALS

Exodus 28:19

And the third row a ligure, an agate, and an amethyst.

Where is amethyst found? In rocks worldwide, such as basalts from Brazil. One exciting place to find amethyst is a geode. It is like a treasure hunt to break open this spherical rock and find lovely crystals of purple inside! Southern Brazil and Uruguay have the largest deposits of amethyst in the world.

What is amethyst used for? Gemstones are used to adorn jewelry and other ornamental objects. Amethyst was one of the gemstones on the "breastplate of justice," part of the ephod that Aaron was to wear as high priest (Exodus 28:15).

Fun Fact: Amethyst is considered the most precious of the quartz minerals. The most valuable gemstone is deep purple with tinges of violet. Amethyst found in the United States is not considered jewelry grade.

Geodes scattered over the Wadi el–Battikh in Egypt.

GEODES

Geodes are fascinating displays of God's grace. These often hollow structures are created in sedimentary and volcanic rock. The round or sphere-shaped oddities are <u>formed</u> because there are cavities inside them, often left behind from gas bubbles, which water and minerals pass through. The rock exterior hardens and crystals form on the inside.

Geodes are very diverse in color. Many are composed of quartz minerals of a variety of colors: banded milky quartz (white) and jasper (red) geodes are called agate. Some quartz-rich geodes open to reveal amethyst crystals or flinty chalcedony. Some are not quartz at all, but have carbonate minerals like calcite and dolomite.

The Rubens Vase from the Byzantine Empire was carved from a single piece of agate. This extraordinary vase was most likely created in an imperial workshop for a Byzantine emperor. It eventually made its way to France, probably carried off as treasure after the sacking of Constantinople in 1204 during the Fourth Crusade, and in 1619 was purchased by the great Flemish painter Peter Paul Rubens.

In 2 Corinthians 4:3-7, Paul was telling the church at Corinth about the beautiful mystery of God's grace that is hidden within very ordinary vessels:

But if our gospel be hid, it is hid to them that are lost: In whom the god of this world hath blinded the minds of them which believe not, lest the light of the glorious gospel of Christ, who is the image of God, should shine unto them. For we preach not ourselves, but Christ Jesus the Lord; and ourselves your servants for Jesus' sake. For God, who commanded the light to shine out of darkness, hath shined in our hearts, to give the light of the knowledge of the glory of God in the face of Jesus Christ. But we have this treasure in earthen vessels, that the excellency of the power may be of God, and not of us.

While he was not speaking of geodes, the message of Paul made reference to something that looked so ordinary on the outside (a clay vessel), yet held a treasure within. In that time period, people often kept what little treasures they had in common pots that would not attract attention. In a similar way, God takes us and pours His Spirit into us, saving us and forming His beautiful life and power within us. No matter how ordinary one may look on the outside, if they are in Christ, He has made us so much more extraordinary than any geode!

The Uniqueness of Each Mineral

As you learn about the structure, color, shape, and other characteristics of a sampling of the many minerals included in this book, it will help you see just how creative our God is. He has breathed that same creative spirit in each of us. If we are overwhelmed with His creative fingerprint on minerals, we should begin to study the human body and realize that we are His masterpiece! God has given us dominion (rule) over the things that He has provided for us telling Adam and Eve in Genesis how they are to subdue the earth. In other words, make good and wise use of the things He created. Minerals are a precious gift to us. It is our job to find ways they can be used to glorify Him. Think about this as you begin to build your own mineral collection.

COPPER

Native copper (Cu) is a red-brown colored mineral that is often used in construction for its ability to easily conduct electricity. The malachite in copper ore turns copper green.

Native copper

Search for Old Pennies

Look in your house for pennies older than 1983. I just did, and out of ten coins I found only one older than 1983. (It was a 1964. I may keep that one because I was 12 years old then!) After 1982, pennies have actually been copper-plated zinc coins, which are cheaper to produce.

The Statue of Liberty, a symbol of our great nation, was a gift from France in the late 1800s. It is completely covered in copper!

Gutenberg's Revolution

In the 1450s, a German named Johannes Gutenberg invented something that revolutionized our world. He found a way to use copper and other metals to construct a printing press. This machine could print a larger amount of text than had ever been possible before. He was able to print the entire Bible for the first time ever. Was it an accident that this invention called for copper and other metals that Gutenberg had access to? Think of how this one invention made it possible to get God's Word into so many countries, churches, and homes.

Biblical Word Search

Do a word search on *copper* in the Bible. How many times is it used? Is copper (or brass or bronze) mentioned more in the Old Testament than the New? What is the Greek word for copper? The Hebrew? Try using various versions to see if copper is used more often by certain translators.

ROCK SOLID
IT'S IN THE WORD
MINERALS

Iron is taken from the earth,
And copper is smelted from ore.
Job 28:2 (NKJV)

Chemistry of Copper Minerals

How are the words *cuprous* and *cupric* used in chemistry? In the mineral cuprite (Cu_2O), what is the valence of the copper? Is it any different than the valence of the copper atoms in tenorite, a mineral with the formula CuO? On the right is tenorite, the black mineral mixed in with the green and white.

Cuprite　　　　**Tenorite**

What is a Mineral?

How important is asking questions in science? The answer is very important! A good question is critically important to a correct understanding. During our Lord Jesus' several trials that first Easter, "Pilate saith unto him, What is truth?" (John 18:38). Now that is one key question in every endeavor — spiritual or scientific. What is truth? Old Aristotle, the Greek philosopher who lived many years before Jesus' Resurrection, said. "To succeed we must ask the right preliminary questions." C.S. Lewis quotes this in his book *Miracles: A Preliminary Study.*[1]

So here is a question for us: what is a mineral?

For the purpose of this book, we will define a mineral as having five foundations.

1. A mineral is a special group of compounds created by God.
2. A mineral has a fixed chemistry.
3. A mineral contains a crystalline structure.
4. A mineral is inorganic.
5. A mineral exists as a solid.

Blue sapphire

1 A Special Group of Compounds Created by God

The first foundation is the most important one to understand. Most secular definitions of mineral begin with "a naturally occurring chemical compound." From an evolutionary view, a specific mineral, like amethyst, is incidental or naturally occurring because of fixed chemical laws. Beware of terms such as *naturally occurring* or *Mother Earth*, which may write God completely out of the picture! This same reasoning explains that a dinosaur evolved through time, formed by biological laws like natural selection. God's Word teaches us that He created everything, for example dinosaurs and minerals. Just as God spoke the minerals into being, with all their different chemical compositions and structures unique to each kind, so God designed and created the dinosaur kinds.

It is interesting to know that the mineralogists contemporary with Darwin would classify the minerals just as they did animals. For instance, the mineral corundum would be called the mineral *species*. A corundum that was blue (sapphire) would be a *variety.*

1 Lewis, C. S.. *Miracles: A Preliminary Study.* San Francisco: HarperOne, 2001.

2 A Fixed Chemistry

Minerals have a fixed chemistry. Let's look at three minerals. Native gold (Au) consists primarily of gold atoms. If you look at the chemistry of an amethyst, it is fixed around the chemical formula SiO_2. What this means is that for every two oxygen atoms there is one silicon atom in the crystal structure. The same is true of jasper (SiO_2).

Some minerals have far more complicated chemical formulas. Augite $(Ca, Na) (Mg,Fe,Ti,Al) (Al, Si)_2 O_6$ is one example. This is the most common pyroxene mineral. It is called an aluminosilicate. However, it can have five other elements in it: calcium, sodium, magnesium, iron, and titanium. Even though the formula for augite is more complicated than that of amethyst, it is still fixed.

Obsidian

3 A Crystalline Structure

There are six basic types of crystal structure. To a mineralogist, these six categories are called crystal systems: cubic, tetragonal, orthorhombic, hexagonal, monoclinic, and triclinic. Some mineralogists add a seventh system called trigonal, but this can be viewed as a subset of hexagonal. I have come to believe that the six crystal systems are an active reminder of the six days of creation.

4 Inorganic

This means they do not contain any plant or animal materials. Now they may contain the element carbon, but not as C_6H_6 (benzene), but rather as CO_3 (carbonate).

5 Exist as a Solid Form of Matter

You can heat up certain minerals at high temperatures and they become molten or liquid. But they do not remain in a liquid state after the heat source is removed.

What Minerals are Not!

Sometimes it is easier to understand what something IS by knowing what it is NOT. As you study minerals, some are not easily identified because of a particular characteristic they have. The gemstone jet looks a lot like obsidian (SiO_2). Just like obsidian, jet is black and very hard. With a little detective work, we find that it is actually coal, which is organic (full of carbon).

Jet (lignite)

Minerals Are Not:

1. Organic (contains no plants)
2. A fossil (is not the remains of plants and animals)
3. Commonly a liquid
4. A gas (minerals are not invisible particles)
5. A rock (rocks are made up of minerals)

Finding Minerals in Rocks

We use minerals to identify and classify the many kinds of rocks found on our planet. This helps a geologist map rock formations in the field, which is essential in understanding the geology of a region and in exploring for mineral resources.

Minerals make up a rock just as bricks make up a brick wall, in a great variety of arrangements. In coarse-grained rocks the minerals are large enough to be seen with the naked eye. In some rocks the minerals can be seen to have crystal faces, smooth planes bounded by sharp edges; in others, such as a typical sandstone, minerals are in the form of fragments without faces. In fine-grained rocks, the individual mineral grains are so small that they can be seen only with a powerful magnifying glass, the hand lens that a geologist carries. Some are so small that a microscope is needed to make them out. (*Earth*, Press and Siever, 1986 p.52).

Cinnabar

Temperatures Rising

Glass tubes containing the metallic element mercury (used to determine the temperature) have been used in thermometers since their invention in 1714 by Daniel Fahrenheit. As the temperature rises, the mercury expands and rises as well within the tube. Mercury itself is liquid, the only common liquid mineral and is often found in cinnabar deposits.

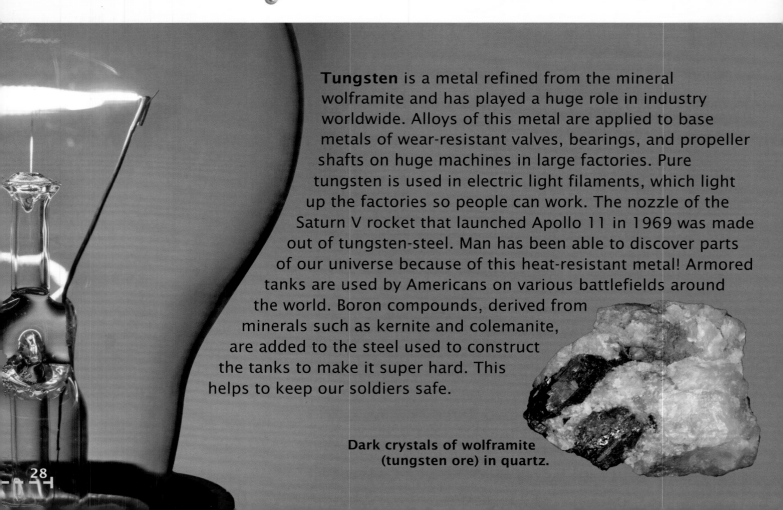

Tungsten is a metal refined from the mineral wolframite and has played a huge role in industry worldwide. Alloys of this metal are applied to base metals of wear-resistant valves, bearings, and propeller shafts on huge machines in large factories. Pure tungsten is used in electric light filaments, which light up the factories so people can work. The nozzle of the Saturn V rocket that launched Apollo 11 in 1969 was made out of tungsten-steel. Man has been able to discover parts of our universe because of this heat-resistant metal! Armored tanks are used by Americans on various battlefields around the world. Boron compounds, derived from minerals such as kernite and colemanite, are added to the steel used to construct the tanks to make it super hard. This helps to keep our soldiers safe.

Dark crystals of wolframite (tungsten ore) in quartz.

From the Garden to Today

Minerals are important to the economy of the world. Civilization has depended on mineral resources from the beginning. In the garden, God provided Adam and Eve with everything they needed, including the salt their bodies would need to function.

From the time that God created Adam and Eve, God gave man the knowledge to use minerals. As Adam and Eve surely marveled at all the minerals He created for them to use, we should be in awe of a Creator who has from the beginning provided these gifts of grace. "But my God shall supply all your need according to his riches in glory by Christ Jesus" (Philippians 4:19). *Christ Jesus* is that Creator, who spoke the universe into existence, with all its rich mineral resources.

CREATOR
SAVIOR
LORD

3

How Do I
Identify *a Mineral?*

This is a book about minerals, not rocks, though rocks are made up of minerals. God has given us science to allow us to classify and organize minerals. The part of geology that deals with the science of minerals is called *mineralogy*. To say we are going to learn to identify a mineral is the same as saying we are going to key them out.

Mineralogists are not the only scientists to use a key or reference chart. Entomologists, who study insects, and ornithologists, who study birds, are just two biological scientists who use keys to help them to identify insects and birds. Ornithologists have a key that includes things to look for, including bird size, bill shape, shape and posture of bird, behavior and habitat of bird, field marks, and even bird voices! The one good thing about identifying minerals is that they don't hop or fly away before you get a good look!

Terms

Luster – *Related to the amount of shine or reflective quality of a particular object.*

Heft – *Related to the weight or heaviness of an object, or testing the weight by lifting it.*

Friedrich Mohs (1773—1839) – *Developed the Mohs' scale to help determine the hardness of minerals.*

Symmetry – *The balance of opposite sides of an object in regard to size and form.*

Level 1   Level 2   Level 3

MINERAL FOCUS	Biotite mica
CHEMICAL FORMULA	$K(Mg,Fe)_3(AISi_3O_{10})(OH)_2$
CRYSTAL SYSTEM	Monoclinic
HARDNESS	2½ – 3
LUSTER	Vitreous to sub-metallic
STREAK	White

What is Biotite used for? It is used to make insulators for electronic components called *capacitors*. It was part of your Grandpa's radio!

Where is Biotite found? In granites, worldwide. Pike's Peak in Colorado (shown above) is one location.

Now I have prepared with all my might for the house of my God the gold for things to be made of gold, and the silver for things of silver, and the brass for things of brass, the iron for things of iron, and wood for things of wood; onyx stones, and stones to be set, glistering stones, and of divers colours, and all manner of precious stones, and marble stones in abundance.

1 Chronicles 29:2

Fun Fact: It is mica that makes granite sparkle.

Beginning the Process of Mineral Identification

The scientific classification is based on chemistry. The amethyst we have spoken of is a variety of quartz (SiO_2), a silicate. The halite (rock salt) we have discussed is a chloride (NaCl). But when we go to the beach and scoop up sand, we do not have a machine under the beach umbrella to determine the chemistry. There must be a visual way to classify, or key out, common minerals. We will start with amethyst and halite and go from there. Remember that amethyst is purple quartz. Quartz is the mineral name; amethyst is the purple variety.

Amethyst

Amazingly Rare Minerals in Comparison to All Known Chemicals

The number of chemicals on earth is 50 million (American Chemical Society, 2009), most of which are the organic chemicals. Of these, about 65,000 are used by industries around the world in making plastics, pesticides, and other so-called industrial chemicals. Around 100 chemicals are actually poisons, necessary for weed and pest control because of the Fall of the created world. As a counter-example, probably 20,000 of these chemicals are used in medicines and drugs that God allowed us to design in order to help our bodies counteract disease. All of us have been sick with at least a headache in our lives, and we can understand what a gift of grace an aspirin can be. Now, where do minerals fit into this big number? There are around 5,000 compounds that fit the definition of a mineral used in this book. Think of that! Do the math — 5,000 out of 50,000,000 = only 1 in 10,000 chemicals is a God-created (naturally occurring) crystalline solid. Of the 5,000 minerals, probably only 100 are used as semi-precious gems, and around a dozen are regarded as precious stones. In the 21st century, four of these minerals are now called precious stones: diamond, emerald, ruby, and sapphire. But we must not miss the Bible blessing in the middle of all these numbers. If all 50 million known chemicals were able to crystallize as geometrically regular solids, they would all fall into the six crystal systems. Keep in mind that this is what is called in science a *thought experiment*, because thousands of these chemicals exist as liquids and gases. In the middle of all this complexity, there are six categories. What a remarkable testimony to God's creative ability! He brings order out of extreme complexity (Romans 1).

A tractor sprays an insecticide or fungicide in an orchard of peach trees.

Luster

What does a banana taste like? Quick, tell me! Well . . . it tastes like a . . . banana. There are some things that are hard to put into words. Luster is an example of this. The word *adamantine* means "reflecting like a diamond facet." If you have never seen photos of vitreous, metallic, dull, and earthy lusters, however, it is hard to know all the subtle luster variations. For instance, there is a unique luster called the Schiller effect. This is an optical property of certain feldspars, which produces a unique shimmering effect as you rotate the specimen in bright light.

Common objects can help you understand luster. Pull a penny out of your pocket or purse. If it is new and shiny, then it has a metallic luster. If the penny is old and has lost much of its shine, it has a dull luster. So just with that one observation, you know two examples of luster: metallic and dull. The reason this is so important is that as we identify an unknown mineral our first question will be: Is the luster metallic or non-metallic? *Dull* is one type of non-metallic luster.

Mystery mineral

Birders or Rock Hounds... Both are Amateur Scientists!

In Ken Kaufman's book, *Field Guide to the Birds of North America*[1], he says "Birding or birdwatching. What is the difference? . . . birding implies a somewhat more active, extreme or radical approach." In the 1950s, the hobby of rock collecting was called "rockhounding, and the hobbyist was nick-named "rockhound." Birders and rockhounds are actually amateur scientists. They may not hold a degree in science, but they have honed their identification skills to key out lots of birds or minerals. You may not want to get your degree in ornithology or mineralogy, but you can still be very knowledgeable about birds and minerals. The more we know about God's world, the more we can appreciate and admire His handiwork. This appreciation and admiration can be a way to worship Him!

1 Kaufman, K., 2000, *Field Guide to the Birds of North America*, Houghton Mifflin (Hillstar Editions) pg. 7.

The Colors of Quartz

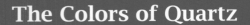

The diverse colors of quartz demonstrate why colors and hues should not be a primary means of mineral identification.

Jasper is opaque chalcedony and is generally red, yellow, or brown.

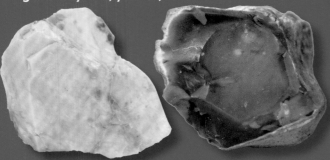

Chert (left) and flint (right) is massive opaque chalcedony, usually white, pale yellow, gray, or black.

Silicified wood is reddish or brown chalcedony.

Agate can be found in a wide variety of colors.

Color

Do not start your mineral ID with color. Ignore the color of the mineral to the last. Remember this one rule: Color defines the value of a gemstone. Color alone cannot be used to identify the mineral.

Although blue is their most common color, sapphires may also be colorless, and are found in many colors including shades of gray and black. This 423-carat Logan sapphire in the National Museum of Natural History, in Washington, DC, is one of the largest faceted gem-quality blue sapphires in existence.

Specific Gravity

When you pick up a handful of salt in your right hand and a handful of sand in your left, which feels heavier? It may surprise you that equal amounts of sand and salt are fairly close in weight or mass. In science class, you may have heard of the terms *mass* and *weight*. These terms are related to the word *heft*.

Pick up a plastic storage bag of sand, first using your biceps (palm upward), and then with your triceps (palm downward). Which feels heavier? Lift the sand both ways. Do you feel the difference? Your bicep muscles are stronger than your tricep muscles.

The specific gravity (SG) of a mineral compares its weight to that of an equal volume of water. Lighter minerals, those with a low SG, are not very dense. Dense minerals, like gold, have a high specific gravity.

Hardness

How hard a mineral is provides yet another clue to its identity. Talc is very soft and diamond is very hard. But how to put a number to it? A German mineralogist, Friedrich Mohs, invented a scale in the early 1800s that is still used today. This scale goes from 1 to 10. The chart to the right lists the minerals from talc (think baby powder) to diamond (think engagement ring). The procedure to see if one mineral, like calcite, would scratch another, like quartz, has been used since the inter-testament times. For example, this test was mentioned by Theophrastus in his treatise *On Stones* in 300 B.C.[1]

How it Works

If calcite (hardness = 3) will scratch quartz, it is harder than quartz. The test is always done both ways, that is, will quartz (hardness = 7) scratch calcite? The answer is, my students, quartz does scratch calcite; therefore, quartz is harder. Absolute hardness is an engineering term for the relative hardness of minerals (like calcite) or materials (like steel). There are 30 different such tests.[2]

For a test in the field, keep these simple rules in mind. Your fingernail is about 2.5 on the Mohs' scale. A U.S. copper penny is almost 3. An ordinary pocketknife is about 5.5, as long as the blade is not some special hardened steel. Carrying a piece of quartz with a pyramid end allows you to test for Mohs' hardness of 7. Be careful to see a true scratch. A hard mineral like corundum, with a Mohs' hardness of 9, will put a metallic powder on the crystal face when a pocketknife is rubbed across it. You are powdering the knife, not the mineral!

Mohs' Hardness	Mineral	Absolute Hardness	Image
1	Talc	1	
2	Gypsum	3	
3	Calcite	9	
4	Fluorite	21	
5	Apatite	48	
6	Feldspar	72	
7	Quartz	100	
8	Topaz	200	
9	Corundum	400	
10	Diamond	1600	

1 Wikipedia, Mohs' scale of mineral hardness, accessed January 5, 2014, http://en.wikipedia.org.
2 McGraw-Hill Dictionary of Scientific and Technical Terms (2nd. Ed.), 1978, p. 723.

What is Scale?

When a research mineralogist takes a photo of a mineral, it should have a scale near the mineral. The scale is often a ruler, in centimeters or inches. Please understand that the scale allows you to see the true size of the quartz crystal or galena cube. One part of all science is measurement. The size of the crystals have real implications regarding creation and the Flood. Very large crystals, feet in diameter, are difficult to explain from an evolutionary viewpoint.[1]

Scale does not have to be a ruler. As a geologist, I would often use my rock hammer for scale when photographing an outcropping of rocks or minerals. Rock hammers are a standard size, and it was always handy to use, since I always have it with me. It would be wise to always have a coin in your pocket, when taking photos of minerals. A U.S. quarter coin is 1 inch (2.5 centimeters) in diameter. When you put a coin beside a mineral before you take a photo of it, you will have documented the size of the mineral. This will be especially helpful when you build a photo collection of minerals you find in museums. With this information, you can compare it to the size of the minerals you later collect in the field.

Something to think about: Museum specimens are usually the best shape and color of a particular mineral. These will be the ones you find photographed in mineral books. If you learn to identify the finest example of a particular mineral or gem, such as its crystal shape and luster, you are training your eye to look for something very close to this when you go to the field. Just as importantly, you will also be better able to spot a fake!

Now think about what you have learned about Christ. Think of His character. He loved His own Father and was willing to leave His home in heaven to do His Father's will in dying for each of us. He spoke truth, even when it meant making enemies. He was compassionate to people who were sick and needed physical healing, like the blind and crippled. He was kind and loving to all people, regardless of age, social status, education, or sinful behavior. Christ is the perfect specimen! He is what we need to examine, so that we will be able to identify Christ in ourselves and in others.

The standard scale used by geologists is a rock hammer. Rock hammers like the one shown above are about one foot in length (12 in or around 30 cm). This photo is from the Mogollon Rim area of Arizona.

1 Lesure, Frank G., 1973, U.S. Mineral Resources: Feldspar: U.S. Geological Survey Professional Paper 820: Washington, DC, U.S. Government Printing Office, p. 220.

Determining the Streak of a Mineral

Because the color of a mineral can vary so dramatically, a streak test can help note a more consistent sample of coloration. One can determine the streak of a mineral by simply scraping it off and collecting this powder on a black unglazed porcelain plate (for lighter minerals) or white unglazed porcelain plate (for darker minerals).

The powdered portion that was scraped off often shows the most reliable impression of color or hue. This is the mineral's streak. Sometimes the streak is identical to the outer color of the mineral, though often it does differ.

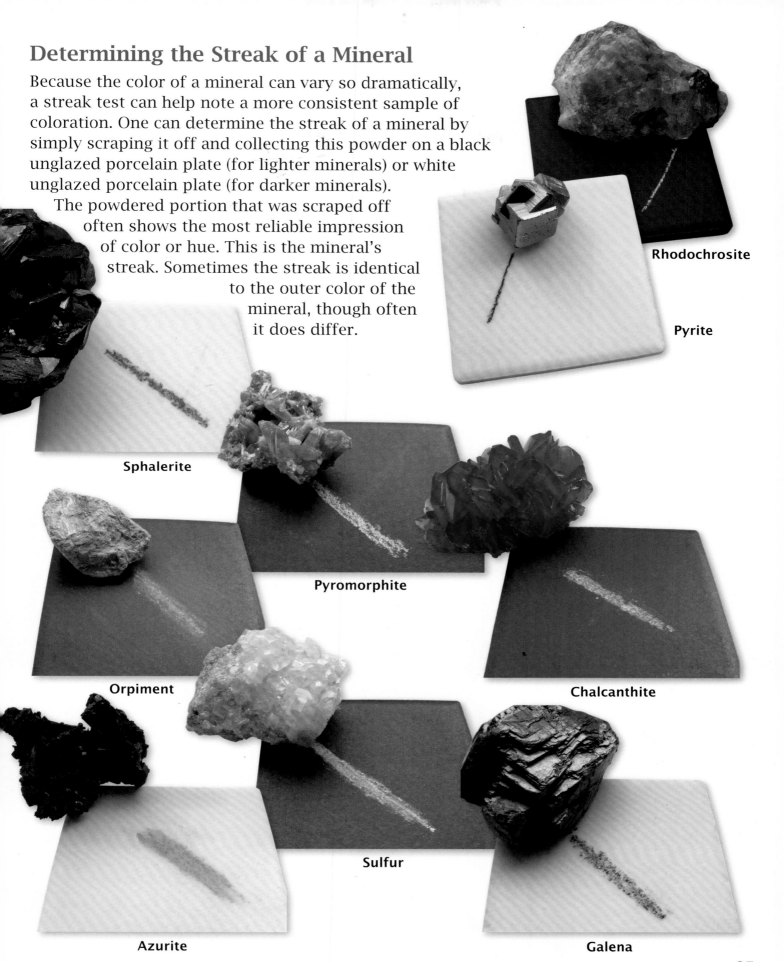

Rhodochrosite

Pyrite

Sphalerite

Pyromorphite

Orpiment

Chalcanthite

Azurite

Sulfur

Galena

Crystals and Symmetry

Now we come to one of my favorite topics, the symmetry we see in crystalline minerals. It is a fact that if we x-ray minerals, no matter how crypto-crystalline (or tiny) they appear to the eye, they have a crystalline signature when magnified. But having a quartz crystal or a big cubic pyrite specimen is so much fun to study.

The following pages show the six crystal systems we will discuss in this book. Some mineralogists use a seventh system called *trigonal*, which is a subset of hexagonal. How do such terms as *cubic, tetragonal, orthorhombic, hexagonal, monoclinic* and *triclinic* hold such clear evidence of God's intelligent design? It is because crystals are a remarkable example of Jesus as Creator, engineer, chemist, and artist.

CUBIC

Cubic system
(as in pyrite, left).
Essential symmetry element: 3 fourfold axis.

Fourfold means the crystal face repeats 4 times (or every 90 degrees) when the crystal is spun on any axis.

ORTHORHOMBIC

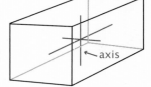

Orthorhombic system
(as in barite, below right).
Essential symmetry element:
  3 twofold axes.

Twofold means the crystal face repeats 2 times (or every 180 degrees) when the crystal is spun on the axis.

TETRAGONAL

Tetragonal system
(as in rutile, below left).
Essential symmetry element:
  1 fourfold axis.

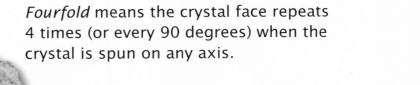

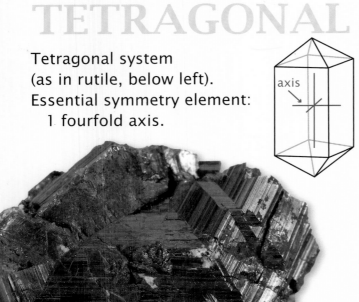

Organic and Inorganic Chemistry

There are thousands of chemicals we use every day. The medicines that have been discovered and manufactured since 1700 now number more than 20,000. In 1853, aspirin was prepared by the French chemist Charles Gerhardt. These chemicals and drugs contain carbon and their chemistry is summarized by the term *organic chemistry*. Because God used these chemicals when He created life in Genesis 1, the foundational chemistry for animals, plants, and humans always include carbon, combined with such atoms as nitrogen, oxygen, and hydrogen. Organic chemistry is often called the *chemistry of life*. The chemistry of minerals is, on the whole, inorganic chemistry. These carbon, oxygen, nitrogen, and hydrogen compounds are replaced by such compounds as silicon dioxide and iron sulfide. This does not mean there is no carbon in minerals. The carbonate minerals such as calcite and dolomite contain calcium and magnesium combined with carbonate ions such as CO_3.

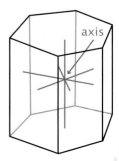

CREATOR

MONOCLINIC

Monoclinic system
(as in orthoclase, left).
Essential symmetry element:
1 twofold axis.

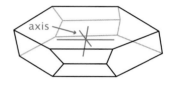

TRICLINIC

Triclinic system (as in lópezite, a rare chromium mineral, below). Not even one axis of symmetry.

Not having a single axis means the crystal face does not repeat when the crystal is spun on the axis.

HEXAGONAL

Hexagonal system (as in emerald, left). Essential symmetry element: 1 sixfold axis.

Sixfold means the crystal face repeats 6 times (or every 60 degrees) when the crystal is spun on the axis.

39

Identifying a Mineral

If you work through each part of the key, it will become clear to you what things you are going to be looking for in order to identify your mineral. By the process of elimination, you will probably find out what your mineral is NOT before you find out what it actually IS!

Take a look at the mystery mineral to the right. Does this mineral look metallic or non-metallic to you? Refer to the chart. Does it reflect the light? Now look at the crystal form. Does it look like one of the crystal forms listed in the basic 6? Cubic, tetragonal, orthorhombic, hexagonal, monoclinic, and triclinic?

Since this is just a photo and not the actual mineral, you can not pick it up and examine the heft, streak, or hardness.

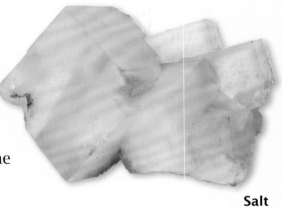

Mystery mineral

Can you see from the photo that the luster is both metallic and shiny? On the right hand side, can you get a hint of a cubic shape? Does the color look like, not golden, but brassy yellow? Now we will give you the answers to the 3 unknown ID characteristics mentioned above.

The streak is black, the hardness is 6, and the heft is lighter than an equal-sized piece of copper. Another way to say this is that it has a specific gravity of 5.0. With all this information, you probably know several things it cannot be. Is it salt? Salt *does* have a cubic crystal structure, but does it have a metallic luster? In the non-metallic grouping of minerals, there is a subgroup called *vitreous*, which means that it appears like broken glass. Salt fits into this category.

Salt

This unknown mineral cannot be salt. Now let's look at still another mineral, stibnite. Stibnite has a metallic luster, so it may be the mineral you are trying to identify. However, when you look at the crystal system, stibnite clearly does not have a cubic crystal system, like the unknown mineral has. The crystals in this stibnite image show that the crystals are long and prismatic. This falls into the crystal system orthorhombic. So you know at least two minerals it is clearly NOT.

The mystery mineral above is actually pyrite!

Stibnite

Mineral Identification Key

While there is no specific order for determining minerals, we have discussed several ways to begin within this chapter:

Luster: Metallic to dull characteristics.
Specific Gravity: Compares a mineral's weight to that of an equal volume of water.
Color: Should never be the primary means of identification.
Hardness: Mohs' scale measures this from 1 (soft) to 10 (hard).
Streak: Scraping a mineral and examining this powder.
Crystal Symmetry: Cubic, tetragonal, orthorhombic, hexagonal, monoclinic or triclinic.

The luster and many other physical properties of a mineral can be checked with a physical examination. Hardness can sometimes be checked by using a coin. Crystal structures can be checked with a magnifying glass, though it's not always easy!

Use the following graph to help assess minerals you may have in your own personal collection, along with the Mineral Identification Guide starting on page 84 of this book.

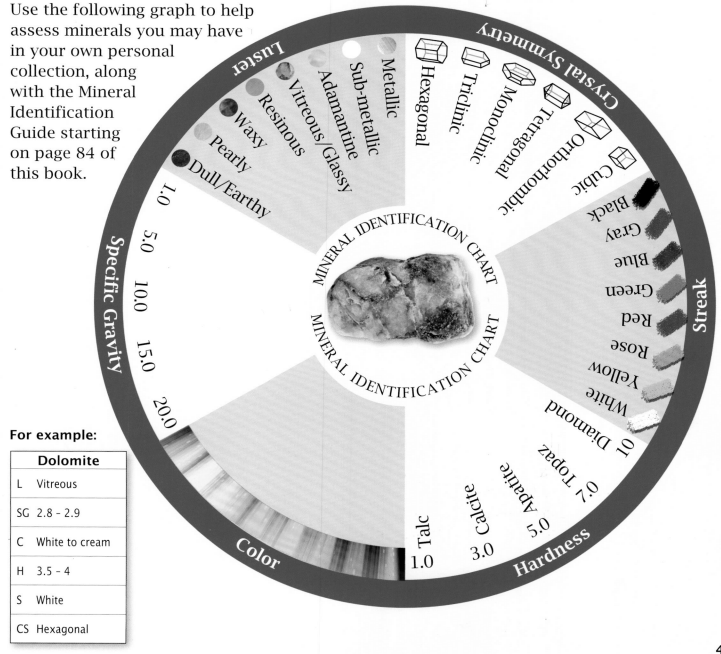

For example:

Dolomite	
L	Vitreous
SG	2.8 – 2.9
C	White to cream
H	3.5 – 4
S	White
CS	Hexagonal

Discovering *the* *Minerals in the Bible*

Bronze-wheeled stand with an animal depiction on the ring and figures in the side panels, which was made in Cyprus, circa 1300 BC.

There are many passages that mention mountains, rocks, minerals, and metals in the Bible. Minerals were created during the Creation week, mentioned in Genesis chapter 1 and 2.

And a river went out of Eden to water the garden; and from thence it was parted, and became into four heads. The name of the first is Pison: ... where there is gold; and the gold of that land is good: there is bdellium and the onyx stone. (Genesis 2:10-12)

The biblical island of Cyprus was known for its copper mining operations some 1,000 years before Christ. This was the period that Cyprus was under Hittite governance. Items from the island were shipped all over the ancient Middle East.

Terms

Geologist – *One who studies the rocks and minerals of Earth, including how they formed.*

Corrupt – *To distort or twist something good into something wicked.*

Analogy – *A comparison of two ideas or things with similar features that help you understand more complex things.*

Atonement – *A reconciling or repairing of a relationship, as done by Christ in His sacrifice so we could stand forgiven before God.*

A mining operation on the island of Cyprus.

Level 1 Level 2 Level 3

MINERAL FOCUS	Jasper
CHEMICAL FORMULA	SiO_2
CRYSTAL SYSTEM	Hexagonal/trigonal
HARDNESS	7
LUSTER	Vitreous
STREAK	White

ROCK SOLID
IT'S IN THE WORD
MINERALS

Revelation 4:3

And he that sat was to look upon like a jasper and a sardine stone: and there was a rainbow round about the throne, in sight like unto an emerald.

Where is jasper found? It is found worldwide. Some localities are known for clear quartz (like Arkansas, US), and others for purple amethyst (like Brazil), but jasper can also be found as crystals and as the crypto-crystalline variety shown here in fossilized logs of the Petrified Forest National Park.

What is Jasper used for? Jasper is a dense material that lends itself to carving. Many specimens are banded and multi-colored. Since ancient times they have been carved to form ornamental buttons, handles, small statues, and all sorts of jewelry. Jasper is one of the stones on the *breastplate of justice* that the priest wore in Exodus 28:20.

Fun Fact: The Egyptians used jasper to carve exquisite artworks. Here is an amulet inscribed with the name of Khaemwaset, son of Ramses II and the priest of the false god Ptah. This is now in the Louvre Museum.

Jasper Creek is the name of a river and a series of cascades and waterfalls in Venezuela, in the eastern sector of Canaima National Park. The water flows over a smooth bedrock of mostly red and black jasper.

Jasper cylinder seal with what appears to be sauropod dinosaurs (Mesopotamia, Uruk Period.)

Jasper goat basket from 19th century Russia.

Hebrew Insights

The Hebrew word translated *onyx* is *shoham*. *Strong's Concordance* defines the word this way: ". . . related to the word blanch, a gem, probably beryl (pale green color): onyx."

To a modern mineralogist, beryl is different than onyx. Beryl is an aluminum-silicate mineral ($Be_3Al_2 Si_6O_{18}$); whereas, onyx is a variety of banded quartz (SiO_2). The "pale green color" of *Strong's Concordance* makes this beryl an emerald. An emerald becomes more valuable as it goes from pale to dark green.

Beryl in the Bible

The color of beryl determines the type of gemstone: pale blue is aquamarine, yellow is heliodor, and pink is morganite. Is it not wonderful how one word of Scripture can open a whole world of possibilities! Should it bother us that modern mineralogy cannot exactly tie down if the onyx of Genesis 2 is SiO_2 or the beryllium aluminosilicate of beryl? Not at all, for God's Word teaches us to expect that there will be some passages hard to understand (the "hard to be understood" of 2 Peter 3:16). But more importantly, did it make any difference to Adam and Eve? No, both onyx (Mohs' hardness of 7) and emerald (beryl = 8) are hard minerals that would have been of great use after the Garden as abrasives and hammers. The exact meaning of the mineral words and rock words in both Hebrew and Greek are areas needing more study. Think of it as a creation scientist does: in eternity future, perhaps Jesus will allow us to research such mysteries.

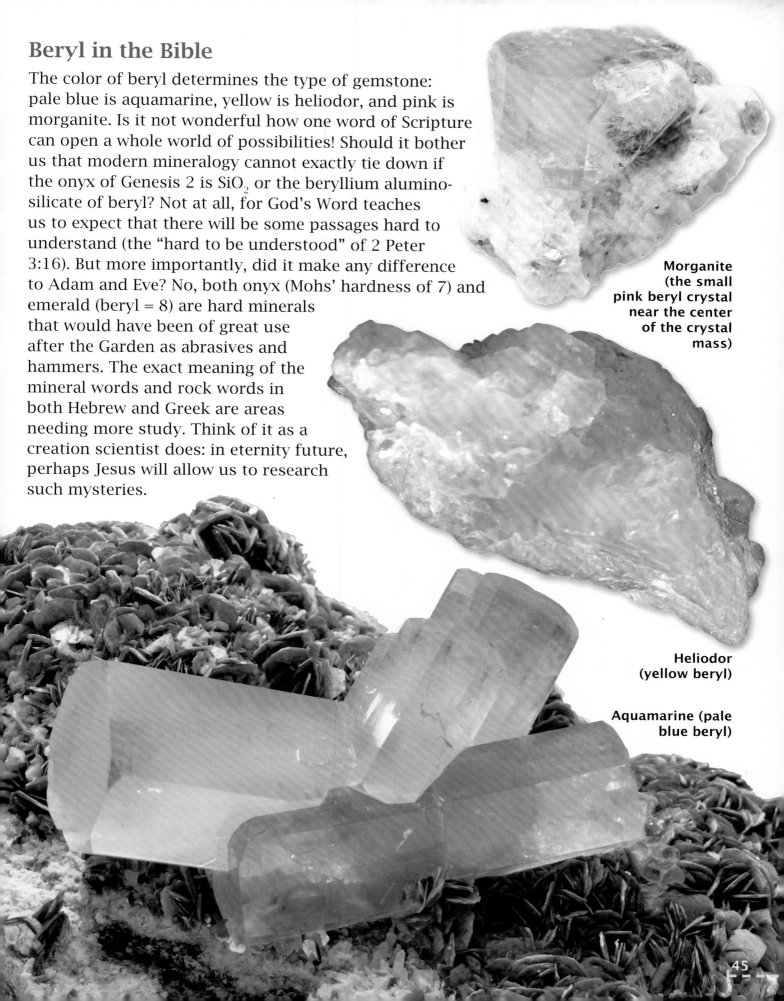

Morganite (the small pink beryl crystal near the center of the crystal mass)

Heliodor (yellow beryl)

Aquamarine (pale blue beryl)

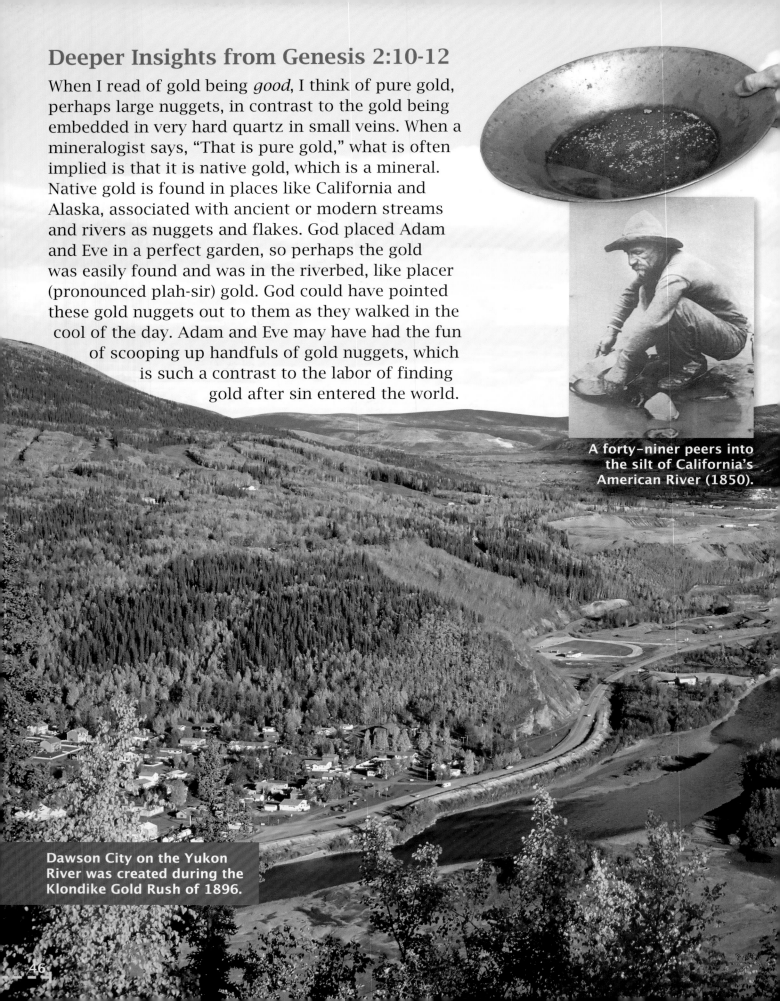

Deeper Insights from Genesis 2:10-12

When I read of gold being *good*, I think of pure gold, perhaps large nuggets, in contrast to the gold being embedded in very hard quartz in small veins. When a mineralogist says, "That is pure gold," what is often implied is that it is native gold, which is a mineral. Native gold is found in places like California and Alaska, associated with ancient or modern streams and rivers as nuggets and flakes. God placed Adam and Eve in a perfect garden, so perhaps the gold was easily found and was in the riverbed, like placer (pronounced plah-sir) gold. God could have pointed these gold nuggets out to them as they walked in the cool of the day. Adam and Eve may have had the fun of scooping up handfuls of gold nuggets, which is such a contrast to the labor of finding gold after sin entered the world.

A forty-niner peers into the silt of California's American River (1850).

Dawson City on the Yukon River was created during the Klondike Gold Rush of 1896.

Gold Rush!

The search for gold, ultimately for quick wealth, spurred on many gold rushes in the 19th and 20th centuries. Though few actually became wealthy from the frenzy of migration that uprooted countless people in pursuit of their dreams, such rushes occurred in Australia, Brazil, Canada, New Zealand, South Africa, and the United States, most notably California starting in 1849, inspiring the name forty-niners. This desire to find a get-rich-quick scheme still distracts and deceives many people today. Remember a "good name is rather to be chosen than great riches" (Proverbs 22:1a).

Gold panner in Grants Pass, Oregon circa 1903.

Miner pouring material from a stream bed into a rocker box, which when rocked back and forth helped to separate gold dust from the sand and gravel.

Sluice boxes of various designs lined up and in use on Nome Beach, Nome, Alaska, 1908.

Miners and prospectors climb the Chilkoot Trail during the Klondike Gold Rush, September 1898.

Minerals Mentioned in Genesis 2 and Ezekiel 28

Genesis 2:10–12

And a river went out of Eden to water the garden; and from thence it was parted, and became into four heads. The name of the first is Pison: that is it which compasseth the whole land of Havilah, where there is gold; And the gold of that land is good: there is bdellium and the onyx stone.

Ezekiel 28:12-13, 17

Son of man, take up a lamentation upon the king of Tyrus, and say unto him, Thus saith the Lord God; Thou sealest up the sum, full of wisdom, and perfect in beauty. Thou hast been in Eden the garden of God; every precious stone was thy covering, the sardius, topaz, and the diamond, the beryl, the onyx, and the jasper, the sapphire, the emerald, and the carbuncle, and gold: the workmanship of thy tabrets and of thy pipes was prepared in thee in the day that thou wast created... Thine heart was lifted up because of thy beauty, thou hast corrupted thy wisdom by reason of thy brightness: I will cast thee to the ground, I will lay thee before kings, that they may behold thee.

Notice that in the passage dealing with Lucifer in the Garden (King of Tyrus), the mineral sardius is mentioned. The modern word for this type of banded brown quartz is *sard*. In the 11th century, a bishop called Marbodius said that if you wore a sardius on a necklace, broach, or ring, it would protect you from the devil's attacks of incantation and sorcery. Marbodius recommended a related sard with white and red bands added to the mix. This mineral is called sardonyx when you have three different colors in bands: white, brown, and red (note Revelation 21:20). What protects us from Satan's attacks? Certainly not a sardonyx necklace! No, we are protected from the Old Serpent by the blood of Christ.

A slice of sardonyx

Tabletop seal of Emperor Alexander II.

The cup of Tolomeo from Alexandria (the Hellenistic period) made of sardonyx.

Sulfur can be collected with ease from fumaroles like this one in Vulcano, Italy, though the process can be dangerous because of the hazardous fumes.

SULFUR

Sulfur is a native element or native mineral. It has been used for thousands of years. Its native element was found in abundance around the world near hot springs and volcanoes, in such places as China, Greece, and India. It has some very odd properties, being both a poison and a medicine, and burning blue though its coloring is bright yellow. It is mentioned in the Bible by the name *brimstone* or *burning stone*.

Sulfur is used in the making of gunpowder, believed to have been invented in China around the ninth century. The Chinese developed a way to combine the mineral sulfur with charcoal and potassium nitrate, thus opening the door for fireworks, for blasting rock in mining operations, and for weapons of war.

The Chinese military used their gunpowder to try and stop the invading Mongol armies, but were eventually overrun. The conquering Mongols then tried to use the technology in their failed invasion of Japan.

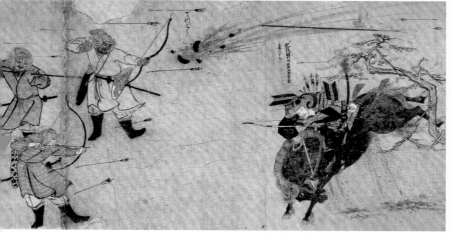

A Mongol bomb thrown against a charging Japanese samurai during the Mongol invasions of Japan after founding the Yuan Dynasty, 1281.

49

The Priests' Clothing of Exodus 28

"And thou shalt make holy garments for Aaron thy brother for glory and for beauty" (Exodus 28:2).

For glory and beauty. Much has been written about the glory of the Old Testament priesthood. As a mineralogist, I see God's glory in the crystal systems represented in the 12 stones of the breastplate (shown on the next page). But we can all see the beauty. From a distance, the breastplate would appear predominantly red. What a wonderful mineral analogy to the atonement that the priest and the blood sacrifice represents.

The diagram is taken from Exodus 28:17–20, with the names and order as they appear in the King James Version. Since Hebrew is written from right to left, the first stone mentioned, sardius, is placed in the upper right, not upper left, as we English speakers would think.

On the following page is a picture showing a possible modern name for each gem, with its color. Now a word of caution — if you compare other Bible references, there are discrepancies. Does this mean that we cannot trust the Bible? Do these differences constitute errors in the Bible? Certainly not! What they teach us is that modern mineralogy cannot be sure of the chemistry and color of each gem, because we do not know what the Hebrew and Greek words are referring to in modern geology. Can you see the humility that this brings to us as creation scientists? We are driven to faith in God's Word and an awareness that "our thoughts are not God's thoughts" (Isaiah 55:8).

Now we will move to a consideration of the colors. Amethyst was, in Bible days and today, a purple gem of quartz. An emerald is an emerald because of the green color, and its mineralogy. Emeralds are a variety of beryl. But sapphires can be green and mistaken for emeralds, yet they are a variety of the mineral corundum. Which one Moses and the moving camp may have been given in Egypt or bought along the way we are not told. But of the green color we can be sure.

Bareketh (Hebrew)
Carbuncle (KJV)
White/black/red (Midrash Rabba)
Levi (Tribe)

Pitdah (Hebrew)
Topaz (KJV)
Green (Midrash Rabba)
Simeon (Tribe)

Odem (Hebrew)
Sardius (KJV)
Red (Midrash Rabba)
Reuben (Tribe)

Yahalom (Hebrew)
Diamond (KJV)
White (Midrash Rabba)
Zebulon (Tribe)

Sappir (Hebrew)
Sapphire (KJV)
Blue black (Midrash Rabba)
Issachar (Tribe)

Nophak (Hebrew)
Emerald (KJV)
Sky blue (Midrash Rabba)
Judah (Tribe)

Ahlamah (Hebrew)
Amethyst (KJV)
Wine red (Midrash Rabba)
Gad (Tribe)

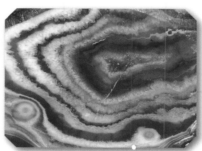

Shebo (Hebrew)
Agate (KJV)
Black/white (Midrash Rabba)
Naphtali (Tribe)

Leshem (Hebrew)
Ligure (KJV)
Blue black (Midrash Rabba)
Dan (Tribe)

Yashpheh (Hebrew)
Jasper (KJV)
Multicolored (Midrash Rabba)
Benjamin (Tribe)

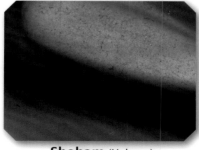

Shoham (Hebrew)
Onyx (KJV)
Black (Midrash Rabba)
Joseph (Tribe)

Tarshish (Hebrew)
Beryl (KJV)
White (Midrash Rabba)
Asher (Tribe)

* The Midrash Rabba is a collection of Jewish rabbinical teachings gathered from the first few centuries after Christ's death and resurrection. Because of varying translations of the minerals, and because minerals often have differing hues, the colors of the 12 stones often vary.

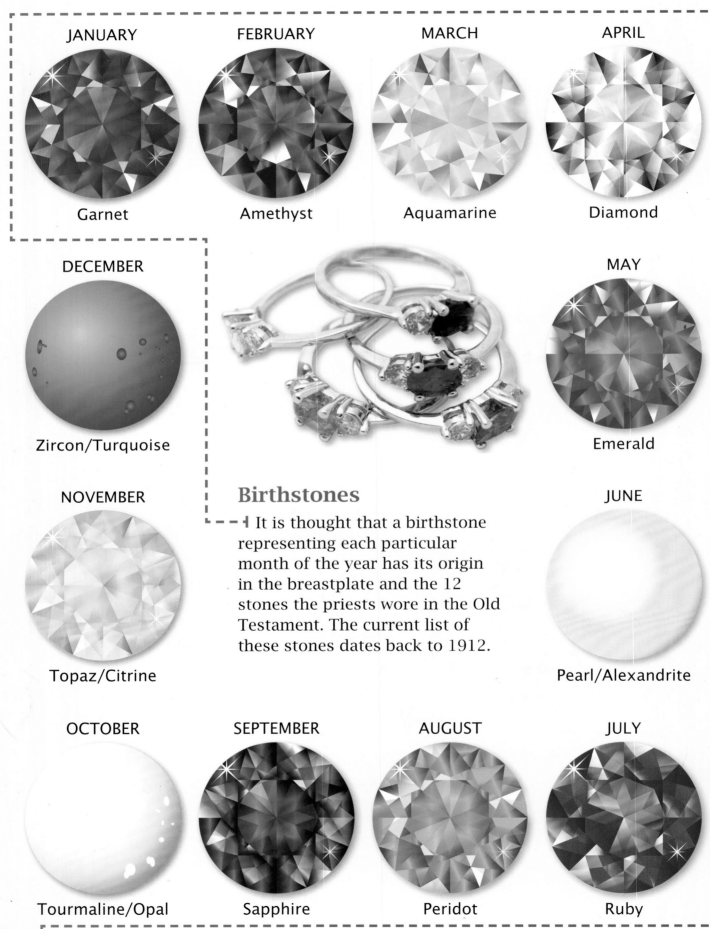

JANUARY
Garnet

FEBRUARY
Amethyst

MARCH
Aquamarine

APRIL
Diamond

DECEMBER
Zircon/Turquoise

MAY
Emerald

NOVEMBER
Topaz/Citrine

JUNE
Pearl/Alexandrite

Birthstones

It is thought that a birthstone representing each particular month of the year has its origin in the breastplate and the 12 stones the priests wore in the Old Testament. The current list of these stones dates back to 1912.

OCTOBER
Tourmaline/Opal

SEPTEMBER
Sapphire

AUGUST
Peridot

JULY
Ruby

Smooth Stones

When young David was preparing to fight the Philistine Goliath (1 Samuel 17), he removed the heavy armor given him, and chose instead five smooth stones from a nearby brook, putting these in his shepherd's bag, and going to face the giant with his stones, sling, and God's strength. These stones were made smooth in the river by its constant movement. You can make your mineral collection smooth as well through the use of a stone polisher or by sandpaper and rock polish. Find what you need in a hobby store or online, and let these smooth, shining stones always remind you that you can fight any giant in your life with God's ever-present strength.

**And I will make thy windows of agates, and thy gates of carbuncles, and all thy borders of pleasant stones.
Isaiah 54:12**

Using Minerals For Good or For Evil

From the beautiful purple amethyst crystals to spectacular zircons, there are about 5,000 crystalline minerals. Do these really have powers, and why is there a resurgence of belief in crystal power? We must be always cautious, for Satan twists God's creation into a lie.

Many followers of the New Age teach that crystals bestow power on the human who possesses them. During January 1999, I had this experience while witnessing online with "Liam2525," who told how he had a "phantom quartz crystal" on his necklace and was intrigued that I was a university-trained mineralogist and a Christian. Liam described himself as a healer who used crystals to "focus his own natural ability." When I asked what power he had through crystals, Liam replied "powers of absorption, the ability to form impressions of people's motives, the power of display, and the ability to form holographic images." Liam commented that crystals could heal and channel (a term used by the occult followers to describe discussions with spirits/demons) and could also serve as a "lens in certain rituals," to determine color associated with energies and vibrations. As a scientist, I was amused at Liam's claims for crystals. However, his reference to contacting the spirits, something that native cultures have traditionally used crystals for, sparked my interest. Is there any biblical evidence that spirits are associated with crystals?

This perfectly ground amulet of rock-crystal was the property of a distinguished woman in eastern Europe in the 6th century.

In Ezekiel 28:13, the angel Lucifer is described as being in "Eden, the garden of God: every precious stone adorned you: ruby, topaz, emerald. . . ." The passage goes on to describe this angel's duties. Amazingly, they revolve around minerals and crystals! In verse 14, this cherub (Satan) is described as walking among "stones of fire." To a geologist, this fits perfectly the rock picture of the formation of crystals such as ruby and emeralds.

In Brazil, there is a famous amethyst locality where the purple quartz formed inside lava tubes. It thus appears that the angel Lucifer, who is also called that old serpent Satan, or the devil, began his existence around beautiful minerals and this beauty was one avenue of pride which led to his rebellion against God. Perhaps it is not surprising, then, that crystals have come to be used in the occult as a doorway to Satan.

The Hamsa, meaning *hand*, is supposed to symbolize the protective hand of the Creator. However, this talisman distorts the initial use of the Hoshen or breastplate stones of the priest, trying to use the precious stones as a way to protect against the evil eye, rather than trusting in God's strength for protection.

Fluoride is added to toothpaste to help make tooth enamel stronger.

The internal pieces of a quartz wristwatch with the quartz crystal oscillator.

White witches, wiccans, and the followers of the New Age believe that crystals have certain powers. This is not a new belief. Pagan and heathen religions of the world, from the time after the Fall of man, share a common idea that crystals have the power to bring happiness or healing. Be warned: Satan does have power. There are demonic spirits in our world who seek to control your life.

Do crystals have any real, God-given purpose? Yes! Many have medicinal uses:

◇ Dolomite (calcium magnesium carbonate) is used as a chewable antacid tablet.

◇ The beautiful fluorite (calcium difluoride) is a source of fluorine which is combined with sodium. This is the agent in toothpaste that prevents tooth decay by strengthening the enamel of your teeth.

◇ Muscovite, a type of transparent mica (potassium alumino-silicate) is a good insulator against heat and electric current. These properties allowed it to be used not only for see-through oven doors, but in early electronics.

◇ In the 1940s through the 1970s, amateur radio was a big hobby (and in some circles still is). Distinct crystals were needed to transmit or receive different frequencies.

◇ Your watch may contain a quartz crystal oscillator. Quartz (silicon dioxide) has an interesting electrical property: if you put a voltage across the crystal, it vibrates at a fixed frequency. Labeled piezoelectricity, it allows very accurate timekeeping.

Whether in medicine or electronics, our Creator God has placed wonderful power in crystals. This power reflects His greater power as Sovereign Creator and Lord. As Romans 1:20 says, "For the invisible things of him from the creation of the world are clearly seen, being understood by the things that are made, even his eternal power and Godhead; so that they are without excuse."

Cello being played with mineral attachment in the microphone.

5 *A World of* Valuable *Minerals*

It is obvious in Scripture that our LORD does not view value in temporal terms, but rather in eternal terms. We cannot buy our way into heaven by donating our silver and gold to the church or any Christian ministry. Yet we do need to remember the passage dealing with gold, silver, precious stones, wood, hay, and stubble (1 Corinthians 3:11–13). Our labor for Jesus is compared to valuable minerals.

The business side of modern life tracks, day by day, the value of gold, silver, copper, and other valuable metals as a measure of world economic health and national worth. Where did these metals come from in the pre-Flood world, and how are they mined today? What minerals were considered valuable in the days of the tabernacle of Exodus, or during Jesus' day as recorded in the Gospels?

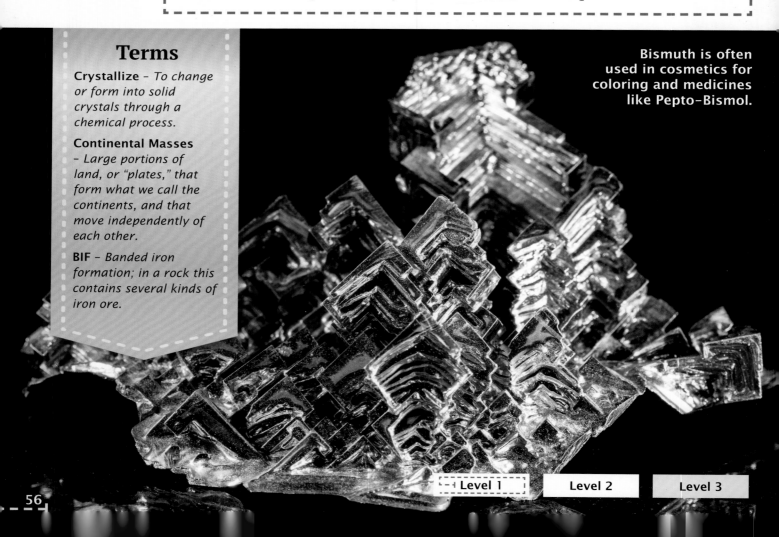

Terms

Crystallize – *To change or form into solid crystals through a chemical process.*

Continental Masses – *Large portions of land, or "plates," that form what we call the continents, and that move independently of each other.*

BIF – *Banded iron formation; in a rock this contains several kinds of iron ore.*

Bismuth is often used in cosmetics for coloring and medicines like Pepto-Bismol.

Level 1   Level 2   Level 3

MINERAL FOCUS	Diamond
CHEMICAL FORMULA	C (Carbon)
CRYSTAL SYSTEM	Cubic
HARDNESS	10
LUSTER	Adamantine
STREAK	Will scratch streak plate

Where is diamond found? The right conditions for diamonds to crystallize occur in the mantle of the earth, generally at a depth of more than 95 miles, below ancient continental masses. Most diamonds come from volcanic rocks. When the magma erupts, it forces diamonds from mantle rocks up to the earth's surface.

What is diamond used for? Gem-quality diamonds are used in jewelry. In fact, one ton in every five tons of diamonds is used for jewelry. However, there are many industrial uses for diamonds.

Jeremiah 17:1

The sin of Judah is written with a pen of iron, and with the point of a diamond: it is graven upon the table of their heart, and upon the horns of your altars;

Fun Fact: In 1477, Archduke Maximillian of Austria commissioned the very first diamond engagement ring on record for his betrothed, Mary of Burgundy. This sparked a trend for diamond rings among European aristocracy and nobility.

The Hope Diamond

The Hope Diamond is the most famous of all the diamonds. It was probably discovered in a mine in India. It is blue and measures 1 inch long, by 4/5 of an inch wide. The size and color make it an exceedingly rare gemstone! And would you believe that it was actually mailed by the jeweler Harry Winston from New York City to the Smithsonian Institution, Washington, DC, in a plain brown mail wrapper? The cost to mail it registered mail was $2.44 for postage, and an extra $142.85 for insurance. (Winston insured it for $1 million.) The postmark is November 8, 1958. Sources vary on the actual value of the Hope Diamond in today's market, but it would be worth far more than the million it was insured for back in 1958. Some estimates are as high as $350 million!

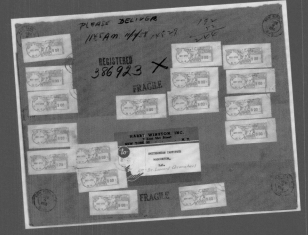

Registered Mail package used to deliver the Hope Diamond to the Smithsonian.

58

Diamonds in Mining and Manufacturing

Most diamonds are mined from steep, conical, pipe-shaped bodies of rocks that form when the hot magma cools. As this volcanic rock erodes, diamonds are often re-deposited by water, and end up in streambeds and oceans. One of the early sources of diamonds was India, then later Brazil. Today, the major diamond reserves are in Botswana, Australia, Russia, Congo, and Angola. South Africa no longer holds the title of producing the most diamonds. Canada has the promise of several diamond mines which have not yet come into full production, and small quantities of diamonds are found in the USA. Crater of Diamonds State Park, in Arkansas, is the only diamond-producing site in the world open to the public.

Although we often think of diamonds as the beautiful gems cut for jewelry, up to 80 percent of diamonds mined are used in industry, since they aren't usable as gemstones. This mineral is ideal in making bearings for instruments in laboratories. Some machines turn at thousands of revolutions a minute. The diamond bearings are so hard and strong that they do not wear down. Diamond cutting tools can slice metal thinner than a human hair. Saws that have diamond-studded edges can cut hard rocks and concrete.

Diamond anvil impactor used to test the hardness of various materials.

The Mir diamond mine, estimated to be the third deepest open-pit mine in the world, is located in the very harsh, cold climate of northern Russia. The winters are so cold around the mine that rubber tires and steel actually shatter, and in the summer months the ground becomes soft and unstable, so the actual time frame that diamonds can be extracted is minimal. These kinds of difficult conditions contribute to the high cost of diamonds and the dangers associated with diamond production.

Blood Diamonds

Diamond is a word that comes from the Greek *adamas*, which meant *unbreakable*. These are the hardest minerals on earth. Diamonds are also very valuable for their intrinsic worth as gifts of love and their other uses as well. They have been sought after for thousands of years.

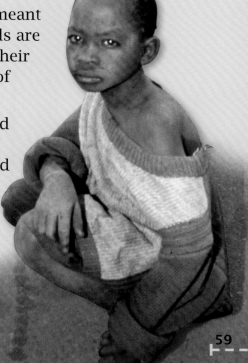

Because many diamond producers come from areas of the world where there are wars or violent conflicts, often helping fund destructive war efforts with their sales, the diamonds uncovered there have come to be called *blood diamonds* or *conflict diamonds*. Such areas of the world to experience this dark side of diamonds include Angola, the Republic of the Congo, and Zimbabwe. Many governments restrict the sale of blood diamonds into their own countries because they do not wish to assist in the destruction that such money produces.

How Diamonds Are Graded: The Four C's

As of May 2018, a one-carat diamond ring ranged in price from $5,000 to $6,000. Less than 900 carats of exceptional diamonds (+D color LC/IF) are sold each year. To give you some idea what "+D color LC/IF" means, consider the four C's of diamonds. Diamonds have unique qualities that are used to assign value to them. Here is how they are actually graded:

Color: Most gem-sized diamonds are rarely perfectly transparent, containing no hue or color. Diamonds of any size that are of the very highest purity are totally colorless, and appear a bright white. If a diamond is affected by chemical impurities which make it yellow, it becomes less valuable. Red diamonds are the rarest.

D-E-F	G-H-I-J	K-L-M	N-O-P-Q-R	S-T-U-V-W-X-Y-Z
Colorless	Near colorless	Faint tint	Light tint, tint is somewhat visible to naked eye	Tinted, Usually yellow to a light shade of brown, shade clearly visible to the naked eye

Carat: The carat is a unit of mass equal to 200 mg (0.007055oz). Many people call this the "weight" of a diamond.

Ct	0.10	0.25	0.50	0.70	1.00	1.25	1.50	1.75	2.00	2.50	3.00
D	3.0	4.1	5.2	5.8	6.5	6.9	7.4	7.8	8.2	8.8	9.4
H	1.8	2.5	3.1	3.5	3.9	4.3	4.5	4.7	4.9	5.3	5.6

Clarity: As a diamond is graded, it is assigned a value on the overall appearance of the stone under ten times magnification. The characteristics of the stone are called inclusions, whereas surface defects are called blemishes. Inclusions which affect a diamond's quality can be clouds, feather, cavities, and crystal or mineral impurities. Blemishes which affect the surface of the gem are nicks, scratches, pits, chips, and polish lines.

 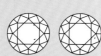 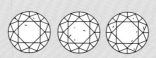 

Internally Flawless IF
Free from internal blemishes visible under 10x magnification (small external details tolerated)

Very Very Slightly included VVS
Inclusions and/or external blemishes very difficult to locate under 10x magnification

Very Slightly included VS
Inclusions and external blemishes difficult to locate under 10x magnification

Slightly included SI
Inclusions and external blemishes easy to locate under 10x magnification

Included I
Medium or large inclusion and/or external blemishes which are usually visible with the naked eye under favorable lighting conditions

Cut: How a diamond is cut greatly impacts its brilliance. It refers to the symmetry, proportioning, and polish of a diamond. A well-cut diamond will be brilliant, and a poorly cut diamond will be less luminous.

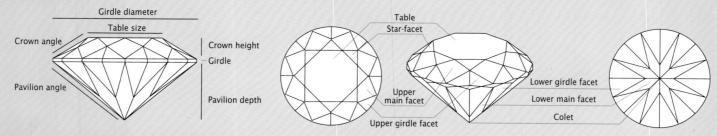

Reference: Wikipedia.org/wiki/diamond_cut

Valuable Minerals for Metals

Sphalerite is the most common ore of zinc, which is used to coat iron and steel so they won't rust.

Hematite is a common ore for iron, which is often made into steel for construction.

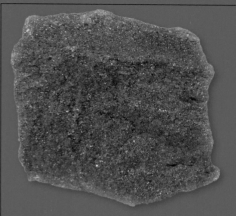

Magnetite is an iron oxide, once used by the Chinese for their compasses and by Romans for their weapons.

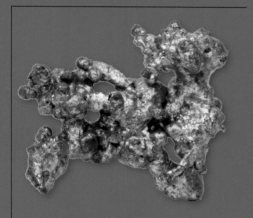

Native copper is rare and is used in electrical wire, as it is an exceptional carrier of electricity.

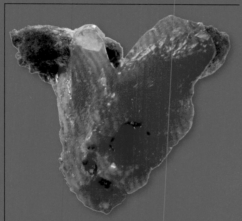

Cuprite is copper oxide, and used as a rare gemstone, even though many crystals are small.

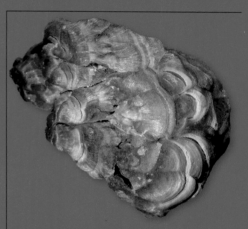

Malachite is a green copper ore mineral, and is used as a decorative stone in construction.

Azurite is a blue copper ore mineral, and is used as a blue coloring called azure blue.

Granites that contain the mineral cassiterite are called *tin granites*, and are used to make pewter.

Arsenic ore/arsenopyrite is a rare ore mineral, and is often found in silver ore veins.

Working with Precious Metals

When considering precious metals, gold immediately comes to mind, along with silver. But where do bronze and brass come from? These metals are smelted from minerals of course! Adam and Eve had access to gold outside the Garden of Eden (see Genesis 2:12). What about silver? Where is silver first mentioned in Scripture? Genesis? Exodus? You look this up. Think of it as homework! Some very large pieces of native silver have been found. Reports from Europe talk about silver nuggets in the tens of pounds. More commonly, silver is combined with sulfur, carbonate, or oxygen. Look at the periodic table in the appendix. Notice where gold and silver are on the table. God designed the chemistry of elements in such a way that we can predict which elements (sulfur – S, oxygen – O, etc.) might combine with Au and Ag.

Native silver, also called a silver nugget, is an ore composed of silver and other elements.

Look up Genesis 4:22, "And Zillah, she also bare Tubalcain, an instructor of every artificer in brass and iron. . . ." Another way to translate this Hebrew phrase is *craftsman in bronze.* So now we have three valuable metals mentioned in the fourth chapter: brass, bronze, and iron. Dr. Morris comments on this passage in his Study Bible (first published in 1995): "Evolutionary archaeologists have attempted to organize human history in terms of various supposed 'ages' — Stone Age, Bronze Age, Iron Age, etc. The Noahic record, however, indicates that early men were very competent in brass and iron metallurgy. . . . Modern archaeology is confirming the high degree of technology associated with the earliest human settlers all over the world."[1]

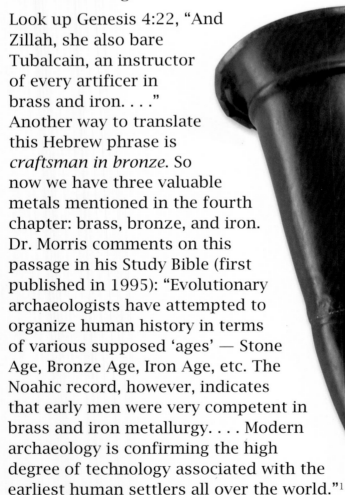

The silver seal of the Hittite ruler Tarkummuwa. This famous bilingual inscription provided the first clues for deciphering Hittite hieroglyphs, and is dated circa 1400 B.C.

A two-handled bronze vessel from the Shang Dynasty (1600–1046 B.C.).

1 Morris, Henry M. *The Henry Morris Study Bible: King James Version.* Revised/Expanded ed. (Green Forest, AR: Master Books), 2012.

Minerals and translations

The American Standard translation of Genesis 4:22 adds another angle to our study of valuable minerals in the Bible. Note this rendering: ". . . the forger of every cutting instrument of brass and iron. . . ." A footnote gives the translation of *brass* as *copper*. Do you see that this translation adds copper to our list of brass, bronze, and iron above? Is this a contradiction in the Bible? Does this make us mineralogists question the accuracy of the translation of Genesis from the original manuscripts? Of course not! We simply cannot be sure of certain metal names, just like we cannot be sure of every mineral name in the ephod breastplate of the high priest. This brings humility to our study of science, and that is a good thing. As Scripture teaches, "Pride goeth before destruction and an haughty spirit before a fall" (Proverbs 16:18).

The "Kang hou gui" from the early Western Zhou (11th century B.C.) is in the British Museum, London.

This shallow silver bowl, or phiale, was used for ritual offerings in Persia around the 6th century B.C.

This beautiful bronze deer is on display at the National Archaeological Museum of Sofia and is about 3,000 years old.

The smelting process has always involved heat in order to extract the desired metal from the ore, and then a chemical process that removes the slag and other elements, leaving behind just the base metal. This basic process has been around for thousands of years, but dramatically modernized with the advent of electricity and large smelting plants.

Valuable Minerals that Provide Copper and Iron

Centered on the states of Arizona, Utah, New Mexico, Montana, and Nevada, as well as the Canadian provinces, you can see copper mines and at a number of locations you can collect the minerals that iron and steel are made from after smelting.

The well-known *Banded Iron Formation (BIF)* rock contains two iron ore minerals: hematite and magnetite. The bottom photo on page 65 shows a modern steel mill. Could Tubal-cain have achieved this level of sophistication? Evolutionary archaeologists, mineralogists, and historians scoff at this suggestion. They argue, "Look at what is involved! Electricity, modern safety rules, huge open-pit mines to give access to the hundreds of tons of hematite and magnetite needed for a mill, and on and on. Could the people of the pre-Flood world have dug a pit like that shown above?" The answer is perhaps. One of Master Books' recent publications presents a detailed argument that "Mankind in Genesis is not simple, unevolved, stupid, or without metallurgy!"[1]

1 Landis, Don. *The Genius of Ancient Man* (Green Forest, AR: Master Books), 2012.

Banded Iron Formation (BIF)

hematite

magnetite

The specimen above shows the banded iron formation, (BIF) containing the minerals hematite and magnetite.

Facing page: Bingham Mine, 2010. It is currently over half a mile deep and 2½ miles wide and was designated a National Historic Landmark in 1966.
Left: A section of the open-pit mining operations of Utah Copper Company at Bingham Canyon, Utah 1942.

The site of the former Bethlehem Steel plant.

The vast, elaborate workings of the interior of a steel plant, including the 'smelting' process done in the blast furnaces.

65

Minerals and Creation *Science*

Geological creationism would predict a certain symmetry in minerals as a result of the creation. The argument from creative design is found in Scripture. For example, see Colossians 1:15-17: "He is the image of the invisible God, the firstborn of all creation. For by him all things were created: things in heaven and on earth, visible and invisible, whether thrones or powers or rulers or authorities; all things were created by him and for him, He is before all things, and in him all things hold together" (NIV). The invisible things, such as atoms and the crystals hidden in a rock, were created by the Lord Jesus Christ. We're told that all things through Him *hold together* or *consist*. This is the case also with crystals. We trace creative design to the Lord Jesus Christ as the Creator, as well as the Savior.

Terms

Radioactive halos – *Concentric circles (circles within circles) left behind by radioactive decay.*

Radioactive atoms *– Atoms that are internally unstable so that they change by radioactive decay at a specific level.*

Radioactive decay – *The breakdown inside an atom that happens so it changes or decays to become stable.*

Black biotite crystals found on Khit Island in Northern Russia

Level 1   Level 2   Level 3

MINERAL FOCUS	Stibnite
CHEMICAL FORMULA	Sb_2S_3
CRYSTAL SYSTEM	Orthorhombic
HARDNESS	2
LUSTER	Metallic
STREAK	Lead-gray to steel-gray

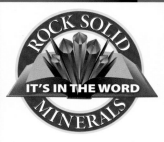

Proverbs 31:30

Favour is deceitful, and beauty is vain: but a woman that feareth the Lord, she shall be praised.

Where can stibnite be found? Stibnite is antimony sulfide and is found associated with quartz, calcite, gold, and other sulfide minerals. It is a widespread mineral, with fine crystals coming from China, Japan, Germany, Russia, Bolivia, Romania, the Czech Republic, Canada, and Mexico.

What is stibnite used for? Egyptians once used stibnite as black eyeliner. The bad thing is that stibnite is an eye irritant! So, beautifully accented eyes gave way to red, tearing ones! More recently, it has been used in lead alloys, mainly for use in batteries, and in semi-conductors for the electronics industry.

Fun Fact: Stibnite can be crushed under a reflecting light microscope, used by geologists to study minerals. As pressure is put on the stibnite, it bends into what is called a *kink band*. The unique part of my research is that I was able to make a videotape of the actual process of the stibnite *kinking*. This type of research could give more evidence for the "invisible things" spoken of in Romans 1:20, and point others to Christ the Creator!

David McQueen, 1979, Kinematics of Experimentally Produced Deformation Bands in Stibnite: Ann Arbor, MS thesis, University of Michigan (published in the journal *Tectonophysics*), 42p.

Granite and Biotite

There is a mineral in the rock called granite that gives us evidence that Genesis 1 is true. This mineral is a type of mica called *biotite*. Radioactive atoms are sometimes trapped in tiny crystals within the structure of biotite. Areas of radioactive damage to the mica give scientists clues that the earth is young, not old.

Biotite minerals in granites have been studied since the 1960s by a creation scientist, Dr. Robert Gentry. The biotite contains a halo that you can see under the microscope. The halo looks like an archery target painted in shades of brown. Gentry's halo research provided evidence that radiometric dating gives a false age of the earth.

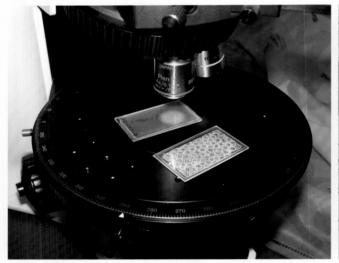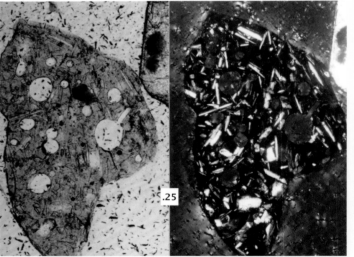

A petrographic microscope (left) with images of a volcanic lithic fragment under plain light (middle) and under crossed–polorized light with filters (right).

Often technology is used to identify and study fine details, either to figure out a sample's composition or structure. One tool that can be used is a petrographic microscope, which helps identify various characteristics of minerals, using a process known as polarized light microscopy. These microscopes can be a little costly, but they are important because they are designed to filter and control light in various ways to study specific aspects of each slice of mineral sample when manipulated by angle and the use of various filters. This technology has its roots in the work of Scottish physicist William Nicol in 1828 with the Nicol prism, which was used in producing polarized light.

Crystals and Creation

A beautiful gemstone is a wonderful testimony to the ordered creation by God. Also, the symmetry of precious minerals is a powerful argument for the truth of biblical creationism, because the mineral world was obviously designed by a wise and organized Creator. Most chemicals that we come in contact with in our daily activities have a crystal structure and symmetry. This is not obvious to us because the crystals of minerals, such as quartz in beach sand or sandstone rock, are so small that they must be magnified or even x-rayed for their symmetry to be seen.

What do we mean by this word *symmetry*? First, the word carries with it the idea of organization. In the case of gemstones or other minerals, this organization is obtained by ordering atoms and compounds at a microscopic level. In many cases we can see this divine organization directly with our eyes because of the symmetry of the crystal faces.

Secondly, the concept of symmetry can be illustrated in God's Word. The first structure that God commanded built, Noah's ark, was symmetrical (see Genesis 6:15–16). Later, Moses was instructed to make a tabernacle. It is interesting to note that the word *pattern* in Exodus 25:9 was used by God to express to Moses the plan, order, and symmetry of this holy structure. The New Jerusalem is described by John using the scientific language of symmetry in Revelation 21:11–21. Note the common association of crystals with God's glory, presence, and plan: "Having the glory of God: and her light was like unto a stone most precious, even like a jasper stone, clear as crystal" (v. 11).

Gypsum

Evolutionists have suggested that the mineral world can be explained on the basis of the operation of time, chance, and impersonal forces. The prominent humanist J. Bronowski, in his book *The Ascent of Man*, said "[crystal symmetry] . . . has been forced on space by matter" (p.174). Do space and matter somehow produce the symmetry of gems, both internally and externally? Of course not! The explanation of an ordered, crystalline, chemical world resides in the Creator who designed the details of the universe.

As mentioned before, mineralogists have identified only six basic types of crystals. This orderly God-created symmetry is a strong argument for our Lord's power. Why is it, with thousands of naturally occurring chemical compounds, we do not see more than six or seven systems? Evolution cannot begin to suggest how many kinds of crystal systems there should be. As an alternative, biblical creationism would predict non-random organization. This is what we see in the mineral world. In this geologist's opinion, we see the six active days or the full seven-day Creation week of Genesis commemorated in the minerals of the very rocks that we walk on.

Modified from author essay: "Crystals and Creation" in *Good News Broadcaster*, Vol. 42, July/August issue, p. 45.

Christian Mineralogists Who Paved The Way

Humphrey Davy (1778–1829) was one of the great chemists of this period. Sir Humphrey was the first to isolate many important chemical elements, to develop the motion theory of heat, to invent the safety lamp, and to demonstrate that diamond is carbon, along with many other vital contributions. He was a Bible-believing Christian, who was highly altruistic and generous. He was also a poet, and for a while, something of a Christian mystic. In his declining years, however, he returned to biblical Christianity and found peace therein.

Davy lamp

William Buckland (1784–1856) was one of the key British geologists in the transition period from biblical catastrophism to uniformitarianism. As a priest in the Church of England, eventually dean of Westminster, he was a Bible-believing Christian. In addition, being trained in the sciences of geology and mineralogy, he became a professor in these disciplines at Oxford. He was a strong creationist and wrote a number of books showing the evidences of design found in these two sciences.

Jean Deluc (1727–1817) was a Swiss naturalist and physicist who studied geology and actually coined the word *geology*. He was strongly committed to the Genesis record of creation and the worldwide Flood. He and his father developed the modern mercury thermometer and the hygrometer. He wrote books on both geology and meteorology and ardently opposed Buffon's evolutionary theories.

Richard Kirwan (1733–1812) was an Irish chemist and mineralogist, president of the Royal Irish Academy for 23 years, and author of the first systematic treatise on mineralogy, also making many contributions to chemistry. He also advocated Flood geology and vigorously opposed the increasingly influential uniformitarian theories of James Hutton.

James Dana (1813–1895) was an American geologist, authoring many influential books on geology and mineralogy. Although he became partly convinced of evolutionism, he continued to believe in biblical Christianity, stating, "The grand old book of God still stands; and this old earth, the more its leaves are turned over and pondered, the more it will sustain and illustrate the sacred Word."

Morris, Henry M. *Men of Science, Men of God: Great Scientists Who Believed the Bible.* (Green Forest, AR: Master Books), 1982.

Radiometric Dating and Radioactive Decay

Often radiometric dating is used to try and prove that rocks are millions of years old. It is all based on radioactive decay, a process in which certain isotopes (or basically a radioactive form of an element) change themselves into different elements. These isotopes are originally unstable and known as *parent elements*. Basically they eject internal particles (protons and neutrons) to achieve more stable forms. These are now atoms of different elements (known as daughter isotopes) than what they were to begin with because they changed their numbers of protons and neutrons. Scientists use about five different forms of parent isotopes to try and date rocks.

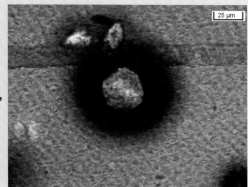

A lot of assumptions can go into this process to date rocks, beginning with the idea that these radioactive atoms have all changed at a constant rate, and that there is a specific number of parent atoms to start with based on how many parent and daughter atoms they find in their sample. They also have to assume that all the daughter atoms were formed through radioactive decay. These assumptions do not take into account forces that could add to or impact the isotopes over the period of their existence.

Image of zircon in a matrix of biotite. The alpha particles emitted by radioactive decay bombard and destroy the matrix biotite, forming what is called a *pleochroic halo*.

Radiohalos, or a colored ring around a small particle of radioactive minerals, are fossilized proof of radioactive decay. Creation scientists use them to prove a biblical timeline rather than the millions of years which secular science uses to date their formation.

The mineral zircon is sometimes an inclusion in the crystal structure of biotite contained within the igneous rock, granite. Dr. Gentry's research paper published during 1974, in the journal *Science,* presented evidence that the halos called into question the radioactive decay rate. This is important because if the radioactive clock has not ticked at the same rate since the creation of earth, all radiometric dating (no matter if it is uranium-to-lead method or the carbon 14 method) is called in question.

Zircon, a mineral that sometimes mimics diamonds in appearance, is a popular form of stone for jewelry. Originally found either colorless or in a variety of colors - red, gold, pink, black, brown, green, and blue. These color changes are a by-product of internal radiation damage in varying degrees that destroy the mineral's crystal structure. This is because of the uranium and thorium within the zircon. This also means that by applying heat to the zircon over time, you can change the color of the crystal. Almost 40 percent of the world's supply of zircon is mined from beach sands in Australia, and it can be found in everything from engines to ceramics, spacecraft and even some ballpoint pens.

A translucent, deep coffee-colored crystal of zircon, which is sharp as a razor and perfectly symmetrical.

7 Minerals and the Lordship of Jesus

An exquisite alabaster perfume jar from the tomb of Tutankhamen.

A woman came to Jesus with an alabaster box of precious ointment or perfume. This would have been made from oriental alabaster, otherwise called calcite alabaster, which was a very valuable substance quarried from near the town of Alabastron in Egypt. It was carved into elaborate perfume bottles, as well as sacred pieces, including sarcophagi for the dead to be buried in. It could even be used in sheets like glass for windows.

The woman proceeded to pour the perfume over Jesus' head, and this caused some of the disciples to get angry, stating that this could have been sold and the money given to the poor. Jesus commended the woman for what she did, and further stated that wherever the gospel was preached, she would be remembered for her act of anointing. He would rise from the dead, Lord and King, and He would make sure this dear woman's act would never be forgotten (Matthew 26:6–13).

Terms

Mineralogy – *A specialist study within geology that focuses on the properties of minerals.*

Crystallography – *The study of crystals, their specific properties and structures.*

Crystallographic symmetry – *Symmetry in crystal systems shows a beautiful balance in their creation.*

Inorganic chemistry – *Chemistry that focuses on non-living chemical interactions.*

Alloy – *A mixture or solution that combines a metal and other metals and elements*

Level 1   Level 2   Level 3

MINERAL FOCUS	Gold
CHEMICAL FORMULA	Au
CRYSTAL SYSTEM	Cubic
HARDNESS	2½–3
LUSTER	Metallic
STREAK	Golden-yellow

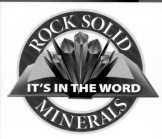

ROCK SOLID
IT'S IN THE WORD
MINERALS

Proverbs 8:10

Receive my instruction, and not silver; and knowledge rather than choice gold.

Where is gold found? Half of the world's gold has been mined in South Africa. South Africa is now only the sixth largest gold producer behind China, Australia, and the United States. Gold appears as veins in igneous rocks, and as dust, flakes, or nuggets in streams and riverbeds.

What is gold used for? The most popular use is in jewelry. Dentists use it to make crowns on teeth to replace damaged enamel. Gold coating on satellite components are vital for temperature control. The Hubble space telescope could not function without the protection from radiation and heat that this gold coating provides.

Fun Fact: The largest gold nugget ever recorded was found in Australia in 1869. It was named "Welcome Stranger," and weighed a whopping 172 pounds. When melted down, it produced about 156 pounds of pure gold!

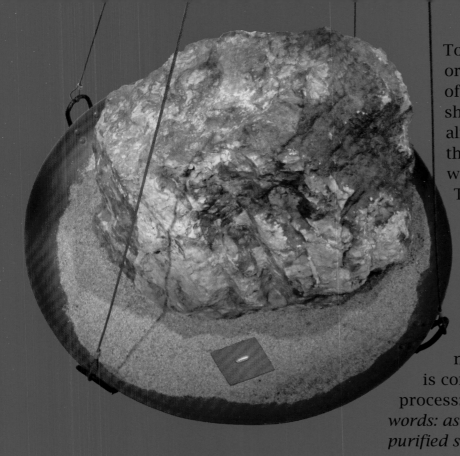

To get enough gold to make a ring or necklace it takes a large block of gold ore. The image on the left shows a piece of gold ore weighing almost 1,895 pounds, as well as the gold that was removed from it weighing just a little over 1 ounce! The process is a lengthy one, which involves crushing the large blocks of ore into smaller pieces, then into a fine powder. Water and chemicals and finally electricity also help remove the gold. At this point, the gold goes through a smelting process, purifying it at nearly 2,100 degrees F! Scripture is compared to a similar purification process: "*The words of the Lord are pure words: as silver tried in a furnace of earth, purified seven times*" (Psalm 12:6).

South American Silver

The Wari civilization, which crossed much of Peru from around 600 or 700 A.D. to 1,000 A.D., produced many gold and silver objects. These artifacts included silver bowls and earrings, and displayed the vast wealth of the capital city, Wari or Huari, in the Andes mountains.

Wari framed mirror of wood and silver.

Gold-embossed Wari ornament in the shape of a bird from Peru.

Standing dignitary of the Wari Empire made from shell and stone inlay and silver.

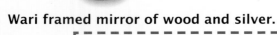

Beauty and the Christian Worldview

As you flip through the pages of this book, what do you see outside of and alongside the text? In the photos of minerals, you see beauty. We have talked about gold as a mineral, a metal, and an alloy. We have explored its possible use in Eden, on the breastplate of the priest, and as a coin in Jesus' day. But why is it pretty, too? Do you see the quandary? Gold is a useful metal because of its durability. But why is it beautiful also? Because our creator God is an engineer and an artist.

Morganite

This is an argument for God's intelligent design that is often not explored. Charles Darwin struggled with function and beauty. He tried to explain why the redness of human blood also allowed his young daughter's cheeks to be rosy. How could time + chance + natural selection = beauty also? Why is amethyst a beautiful purple? Why emerald green? Why the red of jasper? Because our Lord is both an artist and a scientist.

The 7 Cs in the Creation

The best explanation of beauty in the creation is the God revealed to us in the Bible. Is that beauty pervasive and always seen in our fallen world? No. There are no mathematically perfect crystals. We can draw what the hexagonal crystal shape of amethyst should look like. But the actual amethysts from my collection are imperfect. You cannot always see the six-sided symmetry. The actual creation is flawed, fallen, and sinful. Creation was followed by the Curse, leading to the Catastrophic Flood recorded in Genesis. But are these all the Cs in the Bible? By God's grace and mercy, the answer is a resounding NO! As Gary Parker and Ken Ham have pointed out for over 25 years,[1] the 7 Cs are:

Creation, Corruption, Catastrophe, Confusion, Christ, Cross, and Consummation.

In the middle of all these lists and charts, we need to pause. We have argued that you can see God's creative power in minerals. But can we also see His loving personality? His tenderness? His beauty? Ponder this.

1 Ham, Ken. *The Lie: Evolution Revised.* (Green Forest, AR: Master Books), 2012.

MARBLE

Images from the marble quarry at the city of Carrara in Tuscany, Italy. The Carrara marble from here has been fashioned into buildings and artworks since the founding of the Roman Empire. It was utilized in the Pantheon in Rome, as well as the statues of David and the Pietà by Michelangelo during the Renaissance.

Marble craftsmen are able to take a slab of stone and shape it into something beautiful. In Ephesians 2:7-10, Paul wrote: *"That in the ages to come he might shew the exceeding riches of his grace in his kindness toward us through Christ Jesus. For by grace are ye saved through faith; and that not of yourselves: it is the gift of God: Not of works, lest any man should boast. For we are his workmanship, created in Christ Jesus unto good works, which God hath before ordained that we should walk in them."* God sees us as His workmanship or "works of art" that have been created in order to demonstrate good works through our lives. God can take what might seem like empty stone to others and shape us into exactly what we need to be in order to show His love and light to the world.

Building a Mineral
Collection

Putting together a collection of mineral specimens can be very rewarding. There are several ways to begin. You can visit a library, or look up appropriate websites for mineral sets ranging anywhere from $23 to $250.

I think it is much more fun and challenging to put together your own collection from minerals you find in your own backyard or neighborhood. If you live on the beach, start with sand, which consists of quartz. If you live near mountains, you may find outcroppings with copper or galena. If you live in the flatlands, like Louisiana, where hardly any outcrops of rocks containing minerals can be found, travel only a hundred miles to Texas or Arkansas. Here you may find calcite, gypsum, or feldspar. You will be amazed at what you may find in a gravel pit or on a gravel road. Quartz and calcite are usually plentiful in these rocks.

Terms

Mineral specimens – *A sampling of various minerals representing types of mineral categories.*

Classify – *To sort or categorize items so they can be placed in a logical sequence or order.*

Level 1   Level 2   Level 3

All You Need to Get Started

When you begin collecting, you may find large crystals that you enjoy looking at. Or you may like to organize a collection by color — red rose quartz, red garnet, red-brown iron, etc. It is all a matter of personal choice. As you collect more and more mineral specimens, by visiting gift shops in museums, collecting at an outcrop along the highway, or even trading minerals with a friend, you will begin to find different ways to classify them.

Always take a notebook, pencil or pen, a marker, and clear plastic bags that zip or tie closed. When you find a rock or mineral, write in your notebook where and when you found it. Place the rock in the baggie and use your marker (permanent ink) to write a number on the bag. For example, your notebook will read: #1, clear crystal found in a creek just south of Highway 35, near Smalltown. When you get home, carefully take the rock out of the bag labeled No. 1, clean the stone with soapy water and an old toothbrush, and place in a tray or box labeled "#1." Use the keys of hardness, luster, and crystal shape (you may need a hand lens for this), to help determine what mineral it may be. Don't be discouraged if you have trouble identifying your mineral. Even expert mineralogists cannot identify every mineral they find without using more complex tests.

It can be a good investment to purchase a mineral collection, though collecting the minerals yourself can be so much more exciting, as well as educational. This is also a better way to learn how to identify minerals that you might find near your home or at least in your home state. The following tips can help you get started making your very own collection!

◇ Notebook and pencil or pen

◇ Permanent marker (fine point)

◇ Clear plastic bags that zip or tie closed

◇ Light-colored paint or fingernail polish that dries quickly

◇ Geologist's pick or hammer

◇ Magnifying glass or hand lens

◇ Safety goggles to wear when breaking rocks

◇ Work gloves

◇ Backpack to carry large specimens (or lots of small ones)

◇ Tray or boxes with dividers to put your specimens in (an ice cube tray or egg carton are good substitutes.)

◇ Compass (Learn how to use this so you can know what side of the road or creek you collected from. That way, if you want to go back later for more collecting, you can refer to your notebook and find the exact place. It is easy to forget where you collected, especially if you collected from lots of places.)

◇ Bottle of water. This is always important to have so you can quench your thirst. Also, it is fun and helpful to wet a rock so that you can see the mineral better, as the dust washes away.

Cellphone Mineral Collecting

Does your mom or dad have a cellphone that takes pictures? If so, you can begin a cellphone mineral collection without even digging in the ground! Here is how it works: ask your parents or grandparents to take you to a natural history museum. Most major towns like Washington, DC, and New York have museums. If your town has a university in it, there may be a museum on campus. With an adult's help, you can photograph a mineral you like through the display case glass. Many displays have the name of the mineral beside it, but it is my experience that if you focus on the gem, the description will be out of focus. You may have to practice a bit, learning to photograph gems and minerals under glass. You will probably have to turn off your flash, so it will not reflect and ruin your photo. Always carry Post-it notes in your pocket, along with a fine-tipped marker or pen. After you photograph the gem, write down the name of it, the museum name, and date you are taking the photo. Give each mineral a number and write that down on your Post-it note. Now photograph this with your cellphone and it will serve as a description and a reminder of where and when you "collected" it. At a small museum, like those you might find on a college campus, you may find someone who is willing to open up the rock display case and let you have a closer look. The power of a polite question is one thing I taught my own children when they were small. Most people who are passionate about their museum will notice your enthusiasm and will be happy to share information with you. When you show them your cell phone "collection," they will likely be willing to let you handle the minerals, and get great photos!

JADE

Carved Jade from China

For thousands of years, jade has been prized in China for its beautiful coloring, hardness, and ability to be formed into various works of art for both secular and sacred uses. This artwork includes jewelry, intricate figurines, cups, bowls, vases, and large sculptures. Much of jade being mined in China today comes from an area known as Xinjiang in the northwest. Though one often thinks of jade as being a deep green color, and this is the most valuable jade, the colors actually range from white or cream-colored to gray, red, lilac, and near black.

A jadeite sculpture from the Mayan Classic period in Mexico over a thousand years ago.

A jade burial suit from the Han Dynasty (202 B.C. – 220 A.D.), at the Museum of Chinese History, Beijing.

The hilt from this jade dagger with horse head pommel is from India, created during the Mughal dynasty in the 17th century. The jade with carved decoration, inlaid with gold and semi-precious stones.

Jade with dragon designs from China during the Western Han dynasty (4th–2nd century B.C.), Tokyo National Museum.

God's Greater Gifts

A gentle warning . . . God tells us in Matthew 6:19–21: "Lay not up for yourselves treasures upon earth, where moth and rust doth corrupt, and where thieves break through and steal: but lay up for yourselves treasures in heaven, where neither moth nor rust doth corrupt, and where thieves do not break through nor steal. For where your treasure is, there will your heart be also."

As your collection grows, be willing to loan it out to your local or school library, or take it to Sunday school, so others can enjoy it. Remember earlier in the book the story of how a man mailed the Hope diamond to the National Museum of Natural History in Washington, DC? This treasure is seen by hundreds of people each day. What if someone earlier in history had hidden the diamond and later died without letting anyone know where he had hidden it. God does not want us to hoard or hide earthly treasures, but share them so everyone can marvel at His amazing, creative design.

Your collection of minerals can point others to the idea of a creative and artistic God who knew we would love the color, symmetry, and luster of minerals and marvel at how unique each one is. Further study of how these minerals are used help us to understand that they are gifts of grace.

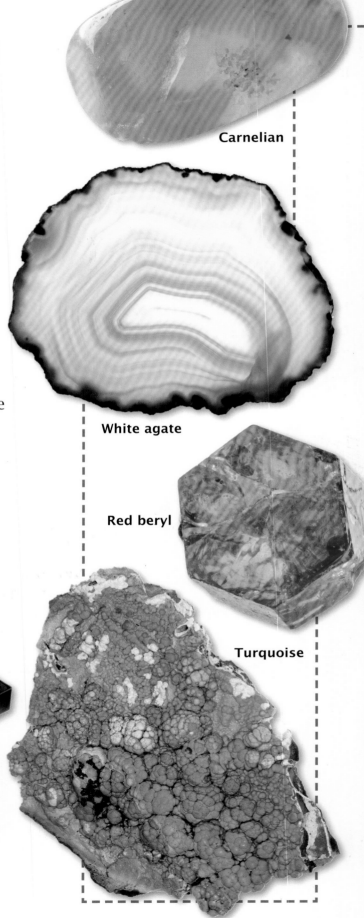

Carnelian

White agate

Red beryl

Turquoise

Beauty and Order Out of Chaos

In Deuteronomy 32:4, we see the biblical analogy between God and rocks: "He is the Rock, his work is perfect: for all his ways are judgment: a God of truth and without iniquity, just and right is he." The Creator is our Rock. In other words, He is our firm foundation for all truth including both science in general, and mineralogy in particular.

In Revelation chapter 21, we see the holy Jerusalem coming down from heaven, with a massive wall and 12 gates, and on the gates are the names of the 12 tribes of the children of Israel, and 12 foundations with the names of the 12 apostles. The building and wall of the city are said to be made of jasper, and the city is of pure gold, with the foundations all made from precious stones. In the King James Version these foundation stones are as follows: jasper, sapphire, chalcedony, emerald, sardonyx, sardius, chrysolyte, beryl, topaz, chrysoprasus, jacinth, and amethyst, and every gate made from a pearl. Some of these magnificent stones are on these pages. How much more wondrous will be this city of God!

Our Father has blessed our modern world with advances in technology, but we could not have high-tech devices without metals such as copper, cobalt, nickel, and iron. These elements are formed naturally in God's world. He gave us all the earth's mineral resources.

The minerals pictured in this book were not created by some undefined designer. The science of mineralogy and crystallography is a remarkable evidence for God's design. Why do nearly 5,000 minerals crystallize in only 6 crystal systems? The same reason that all the animals and plants fall into just a few kingdoms (plants, fungi, animals, etc.). People are not the accidents of biochemistry or organic chemistry, any more than an emerald is an accident of inorganic chemistry.

The value of science is to show an honest student that we have objective, scientific evidence for the creation and the Flood of Noah's day. Since Genesis can be trusted, the Gospel of John can be trusted. May the Holy Spirit convict you today of your need of the Lord Jesus Christ as not only the Creator of amethyst and coming King, who wants to refine you as gold, but also your Savior. The red of jasper reminds us of the Blood of Christ. Have you put your trust in Him?

Blue topaz

Amethyst

Emerald on quartz

Agate

L	Waxy
SG	2.58 – 2.64
C	White to gray, light blue, orange to red, black; banded
H	6.5 – 7
S	White
CS	Hexagonal, but crypto-crystalline

Amethyst

L	Vitreous/gloss
SG	2.65 constant
C	Purple, violet
H	7-lower in impure varieties
S	White
CS	Hexagonal

Aquamarine

L	Vitreous to resinous
SG	Average 2.76
C	Pale blue (variety of beryl)
H	7.5 – 8
S	White
CS	Hexagonal

Augite

L	Vitreous, resinous to dull
SG	3.19 – 3.56
C	Black, brown, greenish, violet-brown; in thin section, colorless to gray
H	5.5 – 6
S	Greenish-white
CS	Monoclinic

Barite

L	Vitreous to resinous
SG	Average 2.76
C	Green, blue, yellow, colorless, pink and others
H	7.5 – 8
S	White
CS	Hexagonal

Beryl

L	Vitreous to resinous
SG	Average 2.76
C	Green, blue, yellow, colorless, pink and others
H	7.5 – 8
S	White
CS	Hexagonal

Biotite

L	Vitreous to pearly
SG	2.7 – 3.1
C	Dark brown to black
H	2.5 – 3.0
S	White
CS	Monoclinic

Calcite

L	Vitreous
SG	2.7
C	Colorless, white
H	3
S	White
CS	Hexagonal

Carbuncle (gem Garnet)

L	Vitreous
SG	3.6
C	Red to orange
H	7 – 7.5
S	White
CS	Cubic

L–Luster SG–Specific Gravity C–Color H–Hardness S–Streak CS–Crystal Symmetry

Colemanite

L	Vitreous to adamantine
SG	2.4
C	Colorless to white
H	4 – 4.5
S	White
CS	Monoclinic

Copper (Native)

L	Metallic
SG	8.9
C	Copper-red to brown
H	2.5 – 3.0
S	Rose
CS	Cubic

Diamond

L	Adamantine
SG	3.4 – 3.5
C	Colorless to white (most common) pinks and blues (very rare)
H	10
S	Will scratch the plate
CS	Cubic

Dolomite

L	Vitreous
SG	2.8 – 2.9
C	White to cream
H	3.5 – 4
S	White
CS	Hexagonal

Emerald

L	Vitreous
SG	2.6 – 3
C	Green
H	7.5 – 8
S	White
CS	Hexagonal

Epidote

L	Vitreous
SG	3.4
C	Pistachio-green or yellow-dog yellow
H	6 – 7
S	Colorless or grayish
CS	Monoclinic

Feldspar (Orthoclase)

L	Vitreous
SG	2.5 – 2.6
C	Cream to orange
H	6 – 6.5
S	White
CS	Monoclinic

Fluorite

L	Vitreous
SG	3.0 – 3.3
C	Most colors, yellow to purple (common)
H	4
S	White
CS	Cubic

Galena

L	Metallic
SG	7.6
C	Lead gray
H	2.5
S	Lead gray
CS	Cubic

Garnet

L	Vitreous
SG	3.6
C	Red is most common
H	7 – 7.5
S	White
CS	Cubic

Gold

L	Metallic
SG	19.3
C	Golden yellow
H	2.5 – 3
S	Golden yellow
CS	Cubic

Graphite

L	Metallic, earthy
SG	2.2
C	Black
H	1
S	Shiny black to steel gray
CS	Hexagonal

Gypsum

L	Subvitreous to pearly
SG	2.3
C	Colorless, white to brown most common, yellow to pink (rare)
H	2
S	White
CS	Monoclinic

Halite (salt)

L	Vitreous
SG	2.1 – 2.6
C	Colorless or white
H	2.5
S	White
CS	Cubic

Heliodor (gem Beryl)

L	Vitreous
SG	2.6 – 2.8
C	Yellow or golden–yellow
H	7.5–8
S	White
CS	Hexagonal

Hematite

L	Metallic to dull
SG	5.3
C	Steel gray
H	5.6
S	Cherry-red to red-brown
CS	Hexagonal

Jasper (Quartz)

L	Vitreous
SG	2.7
C	Red
H	7
S	White
CS	Hexagonal

Kernite

L	Vitreous, silky, dull
SG	1.9
C	Colorless to white
H	2.5
S	White
CS	Monoclinic

Mineral Identification Guide

Kyanite

L	Vitreous
SG	3.6
C	Usually blue but also green
H	4.5 – 6
S	Colorless
CS	Triclinic

Lapis Lazuli (Lazurite)

L	Dull to vitreous
SG	2.4
C	Intense blue
H	5 – 5.5
S	Bright blue
CS	Cubic

Magnetite

L	Metallic
SG	5.2
C	Black, brown-black
H	5.5 – 6
S	Black and magnetic
CS	Cubic

Malachite

L	Adamantine, silky
SG	3.9 – 4
C	Bright green
H	3.5 – 4
S	Pale green
CS	Monoclinic

Morganite (pink Beryl)

L	Vitreous
SG	2.6 – 2.8
C	Pink, peach
H	7.5 – 8
S	White
CS	Hexagonal

Muscovite

L	Vitreous
SG	2.8
C	Colorless, silver-shite, pale green
H	2.5
S	Colorless
CS	Monoclinic

Olivine

L	Vitreous
SG	3.3 – 4.3
C	Green, yellow-dog yellow
H	6.5 – 7
S	White
CS	Orthorhombic

Onyx (Agate)

L	Vitreous
SG	2.7
C	Banded black and white
H	7
S	White
CS	Hexagonal

Pyrite

L	Metallic
SG	5.0
C	Pale brass-yellow
H	6 – 6.5
S	Greenish-black to brownish-black
CS	Cubic

Pyrolusite

L	Metallic, dull, earthy
SG	4.4 – 5.1
C	Black, gray, blue
H	6 – 6.6, 2 when massive
S	Black to blue–black
CS	Tetragonal

Quartz

L	Vitreous – waxy to dull when massive
SG	2.59 – 2.65
C	Can be any color including white and black
H	7
S	White
CS	Hexagonal

Rose Quartz

L	Vitreous as a crystal
SG	2.6 – 2.65
C	Pale pink, rose red
H	7
S	White
CS	Hexagonal

Ruby

L	Vitreous
SG	4.0
C	Ruby red, may vary to purplish or pinkish
H	9.0
S	White
CS	Hexagonal

Sapphire

L	Vitreous
SG	3.95 – 4.03
C	Often blue, but varies
H	9.0
S	White
CS	Trigonal

Sardius (Chalcedony)

L	Vitreous, dull, greasy
SG	2.59 – 2.61
C	Brown, red bands
H	6 – 7
S	White
CS	Hexagonal

Selenite (Gypsum)

L	Pearly
SG	2.3
C	Brown or gray-white, with tints of yellow and green
H	2.0
S	White
CS	Monoclinic

Silver

L	Metallic
SG	10.1 – 11.1
C	Silver, but may tarnish to black
H	2.5 – 3
S	Silver white to black
CS	Cubic

Stibnite

L	Splendent on fresh crystals surfaces, otherwise metallic
SG	4.63
C	Lead-gray, tarnishing blackish or iridescent; in polished section, white
H	2.0
S	Similar to color
CS	Orthorhombic

Mineral Identification Guide

Sulfur

L	Resinous to greasy
SG	2.07
C	Yellow to yellow-brown
H	1.5 – 2.5
S	White
CS	Orthorhombic

Talc

L	Waxlike or pearly
SG	2.58 – 2.83
C	Light to dark green, brown, white, gray
H	1
S	White to pearl black
CS	Monoclinic or triclinic

Topaz

L	Vitreous
SG	3.49 – 3.57
C	Colorless, blue, brown, orange, gray, yellow, green, pink and reddish pink
H	8
S	White
CS	Orthorhombic

Vermiculite

L	Greasy or vitreous
SG	2.4 – 2.7
C	Brown, fulvous, golden yellow, bronze yellow, greenish to blackish
H	1 – 1.5
S	White or yellowish, shiny
CS	Monoclinic

Wolframite

L	Vitreous to greasy
SG	6.1
C	White, yellow, tan, green
H	4.5 – 5
S	White
CS	Tetragonal

Zircon

L	Vitreous to adamantine
SG	4.6 – 4.7
C	Reddish brown, yellow, green, blue, gray, colorless
H	7.5
S	White
CS	Tetragonal

Your Mineral

L	
SG	
C	
H	
S	
CS	

Your Mineral

L	
SG	
C	
H	
S	
CS	

Your Mineral

L	
SG	
C	
H	
S	
CS	

Bible References for Minerals and Metals

The precise identification of some of the terms is unclear, unfortunately, as can be seen by comparing these lists in various translations.

Adamant – Appears in KJV, RSV, REB of Ezekiel 3:9 and Zechariah 7:12. The Hebrew word is sometimes translated *diamond* (Jeremiah 17:1 KJV, NRSV, REB, NAS). It is perhaps best translated *the hardest stone* (Ezekiel 3:9 NIV, NRSV).

Agate – It served on Aaron's breastplate (Exodus 28:19) and by some translations as the third stone on the New Jerusalem foundation (Revelation 21:19 NRSV).

Alabaster – Alabaster may be mentioned once in the Song of Solomon (Exodus 5:15 NRSV, NAS; *marble* in KJV, REB, NIV). In the New Testament (Matthew 26:7; Mark 14:3; Luke 7:37), it refers to containers for precious ointment.

Amethyst – Identical with modern amethyst, a blue-violet form of quartz (Exodus 28:19; Exodus 39:12; Revelation 21:20).

Beryl – Most translations show beryl to be the first stone in the fourth row of the breastplate (Exodus 28:20; Exodus 39:13; REB, *topaz*; NIV, *chrysolite*). The word also occurs in the list of the king of Tyre's jewels (Ezekiel 28:13; RSV, NIV, *chrysolite*; NRSV, *beryl*; REB, *topaz*).

Brass – Brass in the KJV should be rendered *copper* or *bronze*. RSV substitutes *bronze*, retaining brass only in a few places (Leviticus 26:19, Deuteronomy 28:23; Isaiah 48:4; NRSV using *brass* only in Isaiah 48:4).

Bronze – The Bible mentions armor (1 Samuel 17:5-6), shackles (2 Kings 25:7), cymbals (1 Chronicles 15:19), gates (Psalm 107:16; Isaiah 45:2), and idols (Revelation 9:20), as well as other bronze objects.

Brimstone – Refers to *sulfur* (NRSV, NIV). Burning sulfur deposits created extreme heat, molten flows, and noxious fumes, providing a graphic picture of the destruction and suffering of divine judgment (Deuteronomy 29:23; Job 18:15; Psalm 11:6; Isaiah 30:33; Ezekiel 38:22; Luke 17:29).

Carbuncle – In KJV, RSV the third stone of Aaron's breastplate (Exodus 28:17; Exodus 39:10; REB, *green feldspar*; NAS, NRSV *emerald*; TEV, *garnet*; NIV, *beryl*) and material for the gates of the restored Jerusalem (Isaiah 54:12; REB, *garnet*; NIV, *sparkling jewels*.

Carnelian – KJV and sometimes RSV, NASB use *sardius*. A clear to brownish red variety of chalcedony. NRSV reading for one of the stones of the king of Tyre (Ezekiel 28:13; NAS, TEV, NIV, *ruby*; REB, *sardin*) and the sixth stone on the foundation of the new Jerusalem wall (Revelation 21:20).

Chalcedony – An alternate translation for agate as the third stone decorating the New Jerusalem foundation (Revelation 21:19 KJV, NAS, REB, NIV). This cryptocrystalline form of quartz, or silicone dioxide, has many varieties including *agate*, *carnelian*, *chrysoprase*, *flint*, *jasper*, and *onyx*.

Chrysolite – Mentioned in Revelation 21:20, and represents various yellowish minerals. It replaces the KJV rendering *beryl* frequently in the RSV (Ezekiel 1:16; Ezekiel 10:9 ; Ezekiel 28:13) and throughout the NIV but not in NRSV. REB reads, *topaz*.

Chrysoprase/Chrysoprasus – An apple-green variety of chalcedony, the tenth stone of the foundation for the New Jerusalem's wall (Revelation 21:20 KJV).

Copper – Usually alloyed with tin to make bronze which possessed greater strength. The KJV uses *copper* only in Ezra 8:27 (NRSV, NIV *bronze*).

Coral – Calcium carbonate formed by the action of marine animals (Job 28:18; Ezekiel 27:16). NRSV, REB, NAS translated a second word as coral (Lamentations 4:7 KJV, NIV, "rubies").

Crystal – Refers to quartz, the two Hebrew words so translated being related to *ice*. In Job 28:18, KJV has *pearls*; the NIV, *jasper*; but NRSV and NAS, read, *crystal*, while REB has *alabaster*. The glassy sea (Revelation 4:6) and river of life (Revelation 22:1) are compared to crystal.

Diamond – The third stone of the second row of the high priest's breastplate (Exodus 28:18; Exodus 39:11; REB, *jade*; NIV, *emerald*) and one of the jewels of the king of Tyre (Ezekiel 28:13; NRSV, REB, *jasper*; NIV, *emerald*).

Emerald – It is the usual translation of the fourth stone of the high priest's breastplate and one of the stones of the king of Tyre (Exodus 28:18; Exodus 39:11; Ezekiel 28:13; REB, *purple* *garnet*; NAS, NIV, NRSV, *turquoise*). The rainbow around the throne is compared to an emerald (Revelation 4:3).

Gold – Gold occurs in the Bible more frequently than any other metal, being used for jewelry (Exodus 12:35; 1 Timothy 2:9), idols, scepters, worship utensils, and money (Matthew 10:9; Acts 3:6). The New Jerusalem is described as made of gold (Revelation 21:18, Revelation 21:18, 21:21).

Iron – The Canaanites' *chariots of iron* (Joshua 17:16, Joshua 17:16,17:18; Judges 1:19; Judges 4:3) represent a technological advantage over Israel, while the Philistines may have enjoyed an iron-working monopoly (1 Samuel 17:7; 1 Samuel 13:19-21).

Jacinth – A transparent red to brown form of zirconium silicate. It appears in Aaron's breastplate (Exodus 28:19; Exodus 39:11; KJV, *ligure*; REB, TEV, *turquoise*) and the New Jerusalem wall foundation (Revelation 21:20).

Jasper – (Exodus 28:20; Exodus 39:13; Revelation 21:11, Revelation 21:11, 21:18-19) A red, yellow, brown, or green opaque variety of chalcedony. In the RSV for Ezekiel 28:13, jasper translates the word elsewhere rendered *diamond* (REB, *jade*).

Lapis Lazuli – Not a mineral, but a combination of minerals which yields an azure to green-blue stone popular in Egypt for jewelry. It is an alternate translation for *sapphire* (NAS in Ezekiel 28:13; NIV marginal notes).

Lead – A gray metal of extremely high density (Exodus 15:10) used for weights, heavy covers (Zechariah 5:7-8), and plumb lines (compare Amos 7:7-8). Lead is quite pliable and useful for inlays such as lettering in rock (Job 19:24). It was also used in the refining of silver (Jeremiah 6:27-30).

Onyx – Onyx was used on the ephod (Exodus 25:7; Exodus 28:9; Exodus 35:27; Exodus 39:6) and in the high priest's breastplate (Exodus 28:20; Exodus 39:13). It was provided for the settings of the Temple (1 Chronicles 29:2) and was one of the precious stones of the king of Tyre (Ezekiel 28:13).

Pearl – In the New Testament, pearl serves as a simile for the kingdom of God (Matthew 13:46), a metaphor for truth (Matthew 7:6), and a symbol of immodesty (1 Timothy 2:9; Revelation 17:4; Revelation 18:16). Pearl is also material for the gates of the New Jerusalem (Revelation 21:21).

Ruby – The first stone of Aaron's breastplate is sometimes translated *ruby* (Exodus 28:17; Exodus 39:10 NAS, NIV; KJV, RSV, REB *sardius*; NRSV *carnelian*). It also appears as a stone of the king of Tyre (Ezekiel 28:13 NAS, NIV; REB, KJV, *sardius*; NRSV, *carnelian*).

Salt – Used as a seasoning for food (Job 6:6) and offerings (Leviticus 2:13; Ezekiel 43:24). As a preservative, salt was symbolic of covenants (Numbers 18:19; 2 Chronicles 13:5; Matthew 5:13). The *saltpits* of Zephaniah 2:9 were probably located just south of the Dead Sea.

Sapphire – The Hebrew *sappir* is a blue variety of corundum. (Exodus 24:10; Exodus 28:18; Exodus 39:11; Job 28:6, 28:16; Isaiah 54:11; Lamentations 4:7; Ezekiel 1:26; Ezekiel 10:1; Ezekiel 28:13; Revelation 21:19).

Silver – By Solomon's day it was common in Israel (1 Kings 10:27) and was the standard monetary unit, being weighed in shekels, talents, and minas (Genesis 23:15-16; Genesis 37:28; Nehemiah 7:72; Isaiah 7:23), used in idols (Exodus 20:23; Isaiah 40:19), and jewelry (Genesis 24:53; Song of Song of Solomon 1:11).

Soda – *Soda* (Proverbs 25:20 NAS, NIV; Jeremiah 2:22 REB, NIV), or *nitre* (KJV), is probably sodium or potassium carbonate. Other translations prefer *lye* (Jeremiah 2:22 NRSV, NAS). In Proverbs 25:20 the Hebrew text refers to *vinegar* or *lye* or *soda*.

Tin – Sometimes confused with *lead*; articles of pure tin were rare (Numbers 31:22; Ezekiel 22:18, Ezekiel 22:18,22:20). It was principally used in making bronze, an alloy of tin and copper.

Topaz – Second stone of Aaron's breastplate (Exodus 28:17; Exodus 39:10); also mentioned in the wisdom list (Job 28:19) and the list of the king of Tyre's precious stones (Ezekiel 28:13). The ninth decorative stone of the New Jerusalem wall foundation is topaz (Revelation 21:20).

Turquoise – Sky-blue to bluish-green base phosphate of copper and aluminum was mined in the Sinai by the Egyptians and was a highly valued stone in antiquity. Turquoise is sometimes substituted for *emerald* (Exodus 28:18 NAS, NIV); or *jacinth* (Exodus 28:19; Exodus 39:11 REB, TEV).

Source: http://www.studylight.org/dictionaries/hbd/view.cgi?number=T4327

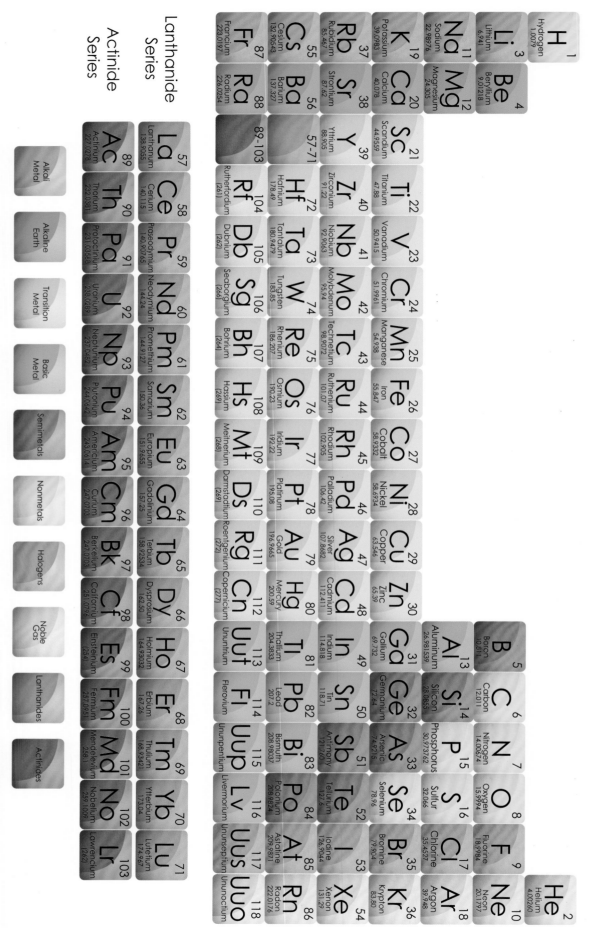

Periodic Table of the Elements

Subject Index

Bible Passages

Old Testament

New Testament

Author Page

Professor David R. McQueen has been a college teacher since 1980 (George Mason University in Virginia). McQueen has taught both full and part time since then at Virginia State University (Petersburg), East Tennessee State University (Johnson City), the Institute for Creation Research's Graduate School (El Cajon, California), and the University of Louisiana at Monroe (ULM). While a graduate student at the University of Michigan (Ann Arbor) in the 1970s, Mr. McQueen held a National Science Foundation (NSF) Graduate Fellowship. During his undergraduate days at the University of Tennessee (Knoxville) he was awarded numerous scholarships, including the one from the oil field service giant, Schlumberger. As a professional geologist for the United States Geological Survey (USGS- HQ in Northern Virginia), McQueen was given a "group cash award" for being part of the "Metallogenic Map of North America" compilation team during 1980. As a professional hydrogeologist for the Louisiana Department of Environmental Quality (LDEQ), Mr. McQueen has received an award, just a few years ago, from the U.S. Environmental Protection Agency (EPA) for his work in supervising the remediation of contaminated groundwater at a major North Louisiana site to LDEQ and EPA standards for the national program called "Ready for Reuse." He is a member of the Geological Society of America and presented a paper on the mineral gypsum at the 2013 Annual Meeting of the Society in Denver, Colorado.

Our Award-Winning *Wonders of Creation Series*

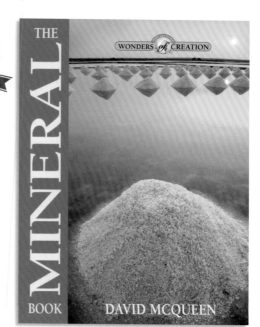

Filled with special features, every exciting title includes over 200 beautiful full-color photos and illustrations, practical hands-on learning experiments, charts, graphs, glossary, and index — it's no wonder these books have become one of our most requested series.

- **The Mineral Book*** reveals the first mention of minerals in the Bible and their value in culture and society.
- **The Ecology Book*** researches the relationship between living organisms and our place in God's wondrous creation.
- **The Archaeology Book*** uncovers ancient history from alphabets to ziggurats.
- **The Cave Book** digs deep into the hidden wonders beneath the surface.
- **The New Astronomy Book*** soars through the solar system separating myth from fact.
- **The Geology Book** provides a tour of the earth's crust pointing out the beauty and the scientific evidences for creation.
- **The Fossil Book** explains everything about fossils while also demonstrating the shortcomings of the evolutionary theory.
- **The New Ocean Book*** explores the depths of the ocean to find the mysteries of the deep.
- **The New Weather Book*** delves into all weather phenomena, including modern questions of supposed climate change.

*This title is color-coded with three educational levels in mind: 5th to 6th grades, 7th to 8th grades, and 9th through 11th grades.

8 1/2 x 11 • Casebound • 96 pages • Full-color interior
ISBN-13: 978-0-89051-802-1

JR. HIGH to HIGH SCHOOL

sample interior from The Archaeology Book

The Ecology Book
ISBN-13: 978-0-89051-701-7

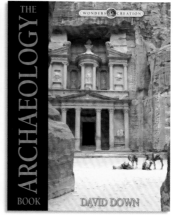

The Archaeology Book
ISBN-13: 978-0-89051-573-0

The New Ocean Book
ISBN-13: 978-0-89051-905-9

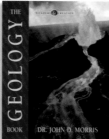

The Geology Book
ISBN-13: 978-0-89051-281-4

The New Weather Book
ISBN-13: 978-0-89051-861-8

The New Astronomy Book
ISBN-13: 978-0-89051-834-2

The Fossil Book
ISBN-13: 978-0-89051-438-2

The Cave Book
ISBN-13: 978-0-89051-496-2